D0773015

INDOOR AIR POLLUTION

RADON, BIOAEROSOLS, & VOC's

Jack G. Kay, PhD.
George E. Keller, PhD.
Jay F. Miller, PhD.

Library of Congress Cataloging-in-Publication Data

Indoor air pollution: radon, bioaerosols, and VOCs/[edited by] Jack G. Kay, George E. Keller, Jack F. Miller.
p. cm.
Proceedings of the Symposium on "Indoor Air Pollution: Its Causes, Its Measurement, and Possible Solutions," presented at the 198th National Meeting of the American Chemical Society held Sept. 10-15, 1989, in Miami Beach, Fla. and sponsored by the society's Division of Industrial and Engineering Chemistry.
Includes bibliographical references and index.
1. Indoor air pollution. 2. Radon — Environmental aspects — Congresses. 3. Organic compounds — Environmental aspects — Congresses. 4. Air — Microbiology — Environmental aspects — Congresses. I. Kay, Jack G. II. Keller, George E. III. Miller, Jack F. IV. Symposium on "Indoor Air Pollution: Its Causes, Its Measurement, and Possible Solutions" (1989: Miami Beach, Fla.) V. American Chemical Society. Division of Industrial and Engineering Chemistry.

TD883.1.I49 1991 628.5′3 — dc20 91-484
ISBN 0-87371-309-5

LEWIS PUBLISHERS, INC.
121 South Main Street, Chelsea, Michigan 48118

PRINTED IN THE UNITED STATES OF AMERICA

Preface

This book is a result of the symposium entitled "Indoor Air Pollution: Its Causes, Its Measurement, and Possible Solutions," presented at the 198th National Meeting of the American Chemical Society, Miami Beach, Florida, September 10 to 15, 1989. The symposium was sponsored by the Division of Industrial and Engineering Chemistry of the American Chemical Society and organized by the three of us.

Some of the chapters included here were added subsequent to the symposium, and some of the papers that were presented orally will be found elsewhere in the professional journals.

We think that the chapters in this book represent the state of the art with regard to fundamental developments in the areas covered. We have tried to present a cohesive grouping of the subjects addressed without attempting a comprehensive coverage of indoor air pollution.

We wish to thank James E. McEvoy and the officers of the Division of Industrial and Engineering Chemistry for inviting us to organize the symposium and for sponsoring this publication. We hope that you, the readers, will find it useful.

Jack G. Kay, George E. Keller, and Jay F. Miller

Contents

Part I

Part II

Part I

CHAPTER 1

Overview of the ACS Symposium on Indoor Air Pollution*

J.F. Miller and G.E. Keller

Over the past few years public interest in indoor air pollution has grown considerably. A large part of this interest began when Legionnaire's disease was discovered in 1977, but the roots of indoor air pollution probably go back to World War II, when energy conservation started, or before. One could argue that the first indoor air pollution problems occurred when man brought fire inside a cave. However, the energy crisis of 1973 certainly exacerbated today's indoor pollution problems. At that time the country began looking for ways to conserve energy. Since the largest user of energy in the country is the building sector, when energy needed to be conserved, that was one of the primary targets. Buildings use even more energy than transportation, consuming 40% of the total energy budget. In order to conserve energy, two types of changes were made. The first was to make the equipment used in buildings more efficient. A modern refrigerator is a considerable improvement over one made 15 years ago. What this means is that if the 125 million refrigerators and freezers now operating in this country were as inefficient as those of 15 years ago, then 20 1000-megawatt power plants would have to be built just to furnish the extra energy. Energy savings allowed by better design of equipment has no repercussions. All the savings have been obtained quite inexpensively.

However, a second type of energy savings was to insulate buildings, stop air leaks, and use as little outside air as possible. This behavior has also saved great amounts of energy, though at a cost. This cost results from not enough outside air entering a building to sweep out pollutants that are generated within the building. Furthermore, pollutants brought in from the outdoors add to those generated inside. Thus, indoor levels of air pollution may be considerably worse than outdoor levels. Indoor air pollution has been termed "the worst health problem in the U.S. accounting for 50% of all illnesses."[2] The cost of the illnesses is staggering, but we cannot turn back energy conservation. If we used as much energy per capita as we did in 1973, we would need 35% more fuel. This is about one half the entire production capacity of OPEC. So, what needs to be done is determine ways in which these great energy savings can be accomplished, while providing clean indoor air.

In order to accomplish such a lofty goal, we must understand what indoor air pollution is and what are its causes. In newspapers, magazines, television, and radio, one is exposed to stories of polluted indoor air. The stories are often just that: stories, bits and pieces of information thrown at the public. Furthermore, in some states the legislators have started acting on these stories. Unfortunately, actions are being taken before it is really known what actions to take, which may be unwise. Consequently, the ACS Symposium was held to investigate questions such as:

- Can we measure biological contamination accurately?
- Can we abate/contain biological contamination?
- Is it possible to stop biological growth by the proper use of bactericides?
- Can we measure volatile organic compounds (VOCs), as well as nonvolatiles, in the air at the low levels typically found in indoor environments?
- What levels of VOCs are acceptable?
- What health risks are associated with low levels of VOCs?
- Are there methods to minimize VOC content in the air?
- Will increasing air intake alleviate indoor air pollution?
- What happens when the outdoor air is polluted?
- Will building bake-out elute VOCs from a building?
- Is building bake-out really a good idea, or is article bake-out prior to installation in the building a better idea?
- What types of air cleaners will give clean indoor air?
- How well will these air cleaners work?

This list does not include radon issues, which is covered in a separate section of the proceedings.

The symposium also investigated reasons that indoor air pollution has obtained national recognition. One driving force is the threat of lawsuits. Lawsuits may be directed at any person involved with the design, construction, maintenance, and ownership of the building. The primary reason for the broad range of defendants is that the causal relationships between indoor air pollutants and human health are not well known. Furthermore, standard techniques for measuring air quality are not well defined. The first article of the proceedings discusses the influence of legal issues in the development of concern for indoor air.

The next portion of the proceedings deals with biological pollution. Different sampling techniques are discussed. Most of the more established sampling methods involve growing samples of microorganisms obtained from the air on nonspecific agar. This sample must be subcultivated so that the species of organisms can be identified. Identification protocols are very time consuming and only a highly trained microbiologist can perform them. Species identification may be quite important because the degree of human response varies over a wide range, depending on the organism. For example, one *tuberculosis* organism can infect a healthy person, while it usually takes about one million *Legionella pneumophila* organisms to infect a healthy person. Also, some organisms, e.g., *Aspergillus niger*, are not usually pathogenic, but can induce allergenic responses in people. Furthermore, organisms like *A. niger* can become pathogenic under certain circumstances. Hence, it is difficult to quickly ascertain whether biological contamination of indoor air may be a cause of indoor air pollution in a particular building.

One quite interesting new assay is an enzyme-linked immunoassay (ELISA) test for *L. pneumophila*. The primary advantage over normal sampling techniques is that the test determines the presence of *Legionella sp*. Furthermore, this test is quite rapid and does not have to involve a trained microbiologist.

The minimization of biological aerosols by two quite different techniques is discussed in two papers. In the first case, the causal agent of a building-related illness was discovered to be fungal growth in the walls. The second paper discusses a sick building sickness that was alleviated by addition of a bactericide to the carpets. This brings up an important point. Sick building syndrome (SBS) is the term applied when at least 20% of the people in a building experience symptoms of illness and the causal agent is unknown. When one or more persons experience illness and the causal agent is known, then it is building-related illness (BRI). A building can exhibit SBS without showing BRI. Usually, however, when a BRI is found, there will be SBS associated with the building. The paper discussing the bactericide is quite interesting in that instead of just trying to solve a sick-building problem, the approach may be used to ensure that a building remains healthy.

The symposium then focuses on chemical air pollution. Sampling techniques are discussed in the first portion. A new solvent extraction technique is investigated. Interestingly, this paper shows how a building with a no-smoking policy can have smoke infiltration from a designated smoking room. A second paper discussing sampling deals with particulate sample techniques and means to obtain data on phthalates and acid levels in the air. The data indicate that levels of phthalates on the order of tens of ng/m^3 can be determined. This extremely low level of detection is quite amazing.

The problematic question that arises is just what levels of VOCs are acceptable. Furthermore, are there specific VOCs that are dangerous at the low levels normally found in buildings? A portion of the question also deals with biological contamination found in indoor air. Molhave[3] recommends that the total VOC level in a building be less than 0.16 mg/m^3 to ensure that inhabitants of a building will not be affected by VOCs. He does allow for levels to be as high as 2 mg/m^3 if nonspecific other factors are not present. These levels may appear quite low, but he recommends such levels to ensure that the building will remain healthy. Unfortunately, this subject was not discussed in the symposium.

One approach to alleviating indoor air pollution is to simply bring more air into a building. In hot summer months, or in the south, the cost of cooling the air can be considerable, but when outdoor air is polluted this approach will not work. A paper looks at how pollution levels of ozone inside a building compare with outside. Ozone has a fairly short half-life, and generally it is assumed to be decomposed to oxygen quite quickly when air enters a building. However, this has been shown to be untrue. Ozone levels inside a building reflect outside levels, especially when the turnover rate is high; thus, when outdoor air is polluted, indoor air will probably be polluted and not a "safe haven".

Building bake-out is a process being investigated as a means to minimize indoor pollutants. The process consists of increasing the temperature inside the building to as high a value as possible, and turning the air over at a high rate to release organics from building materials and furniture. The released chemicals are supposed to then be exhausted so that when the building is cooled and the turnover rate turned down to normal, high levels of VOCs are not produced. However, what one article discusses is that condensation of the released organics may occur on surfaces that are slightly cooler than the desired building temperature. This is because a building cannot maintain a very constant temperature, and thus some surfaces will be cooler. For example, surfaces upon which the sun shines will be hotter than surfaces in the shade.

The same type of effect can be observed in an automobile when the sun shines through the windshield, heating the dashboard. Chemicals, mostly phthalates, are volatilized from the dashboard and then condense onto the somewhat cooler windshield. Furthermore, this article shows that when the temperature is elevated, some heavy VOCs may be eluted from particleboard that otherwise would not have been. Thus, building bake-out may not give all the results desired.

The next question addressed in the symposium is use of air cleaners. The authors discuss the use of standard filters and electrostatic precipitators to remove particulate matter. The electrostatic precipitators appear to work quite well. They finish with a discussion of the inability of activated carbon to adsorb hydrocarbons at normal indoor air pollution levels. It appears that carbon beds would have to be extremely long in order to properly abate indoor air problems.

Where the field of indoor air pollution heads in the 1990s is only partially clear. Some predications can be made with a fair degree of confidence, however. Most probably, better detection methods will be developed. We hope that the latest work with *Escherichia coli* derived monoclonal antibodies will give faster, more accurate tests for microorganisms. This will allow researchers to determine causal links between specific organisms in indoor air and specific human responses. Thus, building owners and consultants will not have to rely on "rules of thumb" when assessing microbial contamination of a building.

Detection methods for VOCs will also become easier to use, more reproducible, and, it is hoped, cheaper. We have great confidence that the 1990s will give faster, more accurate methods, but the cost of analysis may be too high for survey work.

The assignment of risks to specific VOCs is a difficult task at best. Even the risk of a total VOC value may be a difficult answer to obtain. Furthermore, just how the risk relates to indoor air pollution control is nebulous. This is because many papers discuss the relationship of lung responses to ozone for healthy persons, but few papers discuss the same relationship for an asthmatic even though it is widely accepted that asthmatics will respond more acutely to ozone.

Indoor air quality levels must be set such that the pollution level will not harm the most sensitive person, thus protecting everybody in the building. This brings up the concept of the healthy building. The paper discussing use of antimicrobial compounds shows how such compounds can be applied so that microorganisms will not be able to grow on the applied surfaces. This type of application will give assurance to the occupants of the building that their building will be healthy. The work on air cleaners can also be looked upon from that prospective. How can the owner/operator of the building ensure that his or her building will remain healthy? The assurance of building health may allow the owner to rent the building for a higher price and provide the occupants with an atmosphere in which more productive work is performed.

In this regard, we expect to see new anti-pollution techniques and devices on the market in the 1990s. The techniques will include new bake-out procedures to ensure that new materials brought into the building do not exude dangerous chemicals. We will also see better filters, both standard varieties and electrostatic. Also, new types of antimicrobials will be successfully marketed to keep a building clean. Along with these types of actions we expect to see new products coming to the market that will abate VOCs and microbiological aerosols. These new products may be adsorbents, like carbon, which will hold on to VOCs at indoor air pollution levels. Others will burn the VOCs to carbon dioxide and water, e.g., potassium permanganate-loaded catalysts. New equipment for total destruction of all VOCs and microbiologicals will probably be seen in the 1990s. When this type of equipment is available, new regulations specifying indoor air quality levels will probably not be far behind.

The 1990s will be an exciting decade in the understanding of indoor air pollution. Research may unveil causal links between indoor air and personal response, and we may be able to measure and abate pollution to create a healthy building.

The chairs of this symposium thank everybody that was involved with the production of the symposium and its proceedings.

REFERENCES

1. Rosenfield, A.H. and Hafemeister, D., *Sci. Am.*, 258(4) (1988).
2. Massachusetts Commission on Indoor Air Pollution, *JAPCA*, 39(7):924 (1989).
3. Molhave, L., *ASHRAE Trans.*, 92(1A) (1986).

CHAPTER 2

Liability for Indoor Air Pollution

Laurence S. Kirsch, Esq.

Increased knowledge about the possible dangers of indoor air pollution has created a different type of danger for a broad spectrum of the business community: the threat of massive lawsuits. Individuals who are claiming that they have been injured by the air inside their offices or homes have launched lawsuits against employers, manufacturers of chemicals and building products, architects, builders, contractors, and realtors. Even the companies that test and mitigate indoor air face the potential for a liability suit. This article will examine the legal implications of indoor air quality, focusing on the parties that may be liable, the basic liability theories, and litigation to date.

After years of relative obscurity, the subject of indoor air pollution has now gripped the attention of the media and the public. The press and television often report on alleged "sick building" episodes in which various individuals in a single building experience symptoms such as coughing, headaches, respiratory irritation, dizziness, or nausea that appear on entering the building and disappear on leaving.

These media reports have prompted yet further allegations of sick building syndrome. People who believe — rightly or wrongly — that they are being harmed by the building in which they live or work are going to seek out someone to whom they can air their grievances. And unfortunately, many of those least familiar with indoor air pollution issues are on the front line. Building owners and operators are the ones to whom most complaints initially will be directed. They will be forced to evaluate whether these complaints warrant a response, and if so, what kind of response. In making these decisions, they will have to keep in mind that if employees or tenants do not receive what they believe to be a satisfactory response, they are likely to take the next step. In many cases, this will mean calling a lawyer.

Consider, however, the multiple factors that might contribute to sick building symptoms: improper ventilation, bacteria or other microorganisms in a ventilation system, ventilation of automotive pollutants from the outdoors into the building through an air intake duct near the loading duct, chemicals used in the office environment, and materials used in the building itself. This litany of possible causes makes it very difficult for anyone to know *who* is responsible for an indoor air quality problem.

Lawyers for the plaintiff have been creative in devising means of dealing with significant uncertainties in identifying the party or parties responsible for the ills of their clients. When not certain who they should sue, they have sued a large number of possible defendants, constructing novel chains of causation to expand the universe of deep pockets.

A case in point is *Buckley v. Kruger-Benson-Ziemer,*[1] in which a computer programmer brought suit against 9 named and 280 unnamed defendants. The defendants included the building's architects; general contractors; mechanical engineers; heating and air conditioning consulting engineers; manufacturers, distributors, sellers, and installers of the building air conditioning equipment; as well as the manufacturers, distributors, sellers, and installers of carpeting and floor tile and the adhesives used to affix these products to the building; and the manufacturers, sellers, and distributors of "certain chemicals commonly used in offices, including but not limited to toners used in duplicating machines, which chemicals emitted toxic gases into the atmosphere which were harmful to human beings if placed in locations not freely accessible to circulating fresh air".[2]

After what was reported to be a massive amount of discovery, but before trial, various defendants settled the case with Mr. Buckley for close to $700,000.

Other similar cases have been filed in the wake of *Buckley.* In Alaska, for example, employees in an office building housing several state agencies have sued the building owners and operators, along with the building architects, contractors and subcontractors, for sick building problems allegedly resulting from the building's heating, ventilation, and air conditioning ("HVAC") system as well as carpeting used in the building.[3] The plaintiffs were diagnosed as having acute, chronic, and recurrent bronchitis, vertigo, hypokalemis, severe trachea bronchitis, and sinusitis. The plaintiffs also sued a contractor that redesigned and cleaned the HVAC system after the sick building complaints began. This case, based on negligence as well as strict product liability and breach of explicit and implicit warranties, is set for trial in January of 1990.

A slightly different variation of a sick building case is *Stillman v. South Florida Savings & Loan.*[4] In *Stillman,* the savings and loan was a tenant in a building owned by Stillman. When various employees began complaining of sick building symptoms, the savings and loan demanded that Stillman investigate and correct the situation. When Stillman allegedly refused, the savings and loan vacated the premises. Stillman sued for breach of lease, and the savings and loan defended and counterclaimed, alleging that the premises were uninhabitable. The defenses and counterclaims were based on breach of contract, failure of consideration, frustration of purpose, constructive eviction, and negligence. The case is noteworthy as an example of a situation in which an employer may be forced to assert claims based on the alleged injuries of its employees.

Other indoor pollution cases are pending as well. The causes of action in these cases tend to be negligence, breach of express or implied warranties, and strict liability, among others.

For plaintiffs, of course, these lawsuits present formidable barriers. If plaintiffs file suit, it must be within any applicable statute of limitations, they must prove causation, they must prove that a particular defendant caused the injury, and they may need to prove negligence as well. They must also finance the litigation and face the prospect of lengthy delays before receiving any recovery.

For potential targets of such lawsuits, however, these cases are just as troubling. First, defending an indoor pollution case is expensive because it is likely to be complex and require technical expertise. Second, because of the cost of defense, nuisance suits can be brought to coerce settlement. Third, in the event of an adverse verdict, the losses can be staggering.

Most importantly, the prospect of indoor pollution litigation is troubling because the targets have no assurance that they can avoid it. Different juries in different cases can construe similar fact patterns in different ways, leaving companies without any reliable guidance on what they can and cannot do, or must or must not do, within the bounds of their legal obligations.

In view of the expanding number of indoor air pollution lawsuits, companies would be well advised to consider the prospect of liability. By consulting with counsel knowledgeable about indoor air-related issues, it is possible to devise a strategy for minimizing the prospect of suit and for successfully defending any lawsuit that may be brought. In particular, it must be recognized that complaints about indoor air quality have a legal as well as a technical dimension. And in dealing with this legal dimension, as with the technical problems posed by indoor air, an ounce of prevention can be worth a pound of cure.

REFERENCE

1. No. 143393 (Super. Ct. Santa Barbara County, CA, 1987).
2. Complaint, ¶28.
3. *Henley v. The Blomfeld Co.*, No. 3AN-86-10483 (3d Jud. Dist. Anchorage).
4. No. 87-67013 (17th Jud. Cir., Broward County, FL, filed April 27, 1987).

CHAPTER 3

Biological Pollutants in the Indoor Environment

Peter W.H. Binnie

ABSTRACT

Studies carried out by Healthy Buildings International (HBI) found that in over one third of the more than 400 buildings studied, the major pollutants included allergenic fungi, and in greater than two thirds, air supply systems were contaminated with dust, dirt, and microbes. This paper discusses the recognition of sick building syndrome (SBS) as an accepted malady and the possible association of microbial contaminants with SBS. The different types of microbes are described along with the problems they can produce. Sources and spread of microbes within buildings is discussed along with descriptions of methods for sampling from surfaces, water, and indoor air. Special mention is made of sampling for and identification of *Legionella pneumophila,* using a new rapid assay technique, and the importance of correct interpretation of microbial findings against available standards.

INTRODUCTION

In 1982 the World Health Organization[1] made an attempt to define sick building syndrome and, at least since then, it has been recognized as a malady affecting a proportion of people in certain problem buildings. By its nature, SBS is difficult to define, and despite an ever-increasing number of publications on the subject, no single cause has been identified.[2] The indicators that a building may be sick usually are increased staff complaints of minor health symptoms and of stuffy air, intermittent odors, and visible increases in dust levels. Other factors may be uneven temperature zones, noticeable smoke accumulation, and dirt coming out of air supply diffusers. Management may also be aware of increased staff absenteeism and reduced productivity. The symptoms that affected people usually complain of are groups of almost trivial problems such as eye, nose, and throat irritation; headache, rhinitis, and sinusitis with skin irritation; cough, shortness of breath, and general lassitude; and dizziness, nausea, and mental confusion. One common factor often associated with these nebulous groups of symptoms is that they occur on entering the problem building and usually clear shortly after leaving it; another is that they are felt least in the morning and tend to become worse in the afternoon.

The symptoms associated with SBS are quite different from another set that occurs with what has come to be called building-related diseases, although there may be overlaps between them. Building-related disease symptoms, however, usually comprise much more definite and serious symptoms with recognizable medical signs and positive laboratory findings, with the causative agent being much more frequently identified. Such illnesses include Legionnaires' disease, humidifier fever, hypersensitivity pneumonitis, asthma and allergic rhinitis, all of which are usually excluded from descriptions of SBS.[2]

Studies carried out by HBI since 1980[3] have found that in more than one third of the buildings studied, one of the major pollutants was fungi, which is known to be allergenic to susceptible people. The heating, ventilating, and air conditioning (HVAC) systems in 70% of these buildings were either grossly or moderately contaminated with dust, dirt, and associated microbes. The airborne pollutants that may be present in the indoor environment comprise a complicated mixture of living and inanimate materials and may be gases or vapors, fibers, dusts, or microbes. The bacteria that have been isolated from indoor environments, both from HVAC internals and other surfaces as well as from air samples, have included a wide range of species. In some cases species known to cause infections are found, but the vast majority of species identified are generally regarded as harmless to humans. In the case of the fungi, many of the types found are saprophytic and a number that are known to cause allergic reactions and also infections in susceptible people are frequently isolated.

The biological pollutants present in the indoor environment come from a variety of sources, some of which are predominantly found outdoors but many of which occur both indoors and outdoors. In terms of routine building studies that include an assessment of biological pollutants, it is impractical to sample and test for all possible ranges of types that might be present. As far as building surveys are concerned, therefore, we look for those types that can be reliably isolated and identified using well tried and tested sampling and laboratory methods. Thus, we sample and check for the presence of a wide range of fungi and bacteria from surfaces and water and in the air of occupied areas of a building. In the case of biological pollutants such as protozoans, viruses, and chlamydia it is possible to isolate them, but they are more difficult and expensive to culture; this would usually only be done under special circumstances, such as might be found when medical evidence strongly indicates that this is required. Microbes or microorganisms are minute particles of living matter that occur in three main forms generally known as viruses, bacteria, and fungi.

VIRUSES

Viruses are the smallest of the group, although there are overlaps in all the definitions with regard to size, shape, content, and activity of this widely varied and enormous group of living things. Viruses range in size from about 0.3 to 0.02 μm. They consist of a core of nucleic acid (either RNA or DNA but not both) within one or two protein coats and do not grow or reproduce but replicate within host cells. Viruses are incapable of producing their own enzymes to break down complex compounds to simpler substances for their nutrition, and they are generally considered to survive only within living cells that may be human or other animal cells, plant cells, or even bacterial or fungal cells. Once in the cell, they change its normal function to that of a virus production factory that will ultimately destroy the cell and liberate many more virions into the host body. Because of this cellular requirement, it is difficult to conceive that the air conditioning system plays a large role in their multiplication and dissemination. This cannot be completely ruled out, however, as it is known that certain viruses are transmitted from person to person by droplets in the air, as with influenza and measles viruses, or in association with materials shed from the body, such as scabs from the pocks produced by the smallpox and chicken pox viruses. It is possible, therefore, that some viral species may survive long enough in the circulating air protected in some way by other particles or cells and be able to infect building users.

BACTERIA

Bacteria range in size from about 5 to 0.5 μm and are single-celled structures that can only be seen singly under the microscope. When they are cultured artificially, however, they form colonies that are composed of millions of single cells and are readily visible to the naked eye. Bacteria are procaryotic, which means that throughout the cell cycle there is no membrane that separates the nuclear material from other cell components. Some species possess flagellae that allow them to swim through liquid media or swarm over moist surfaces. Bacteria occur in almost every environment, particularly in dusty, dirty places inhabited by humans or other animals. Many of the species of bacteria isolated from buildings are harmless and frequently include members of the genera *Bacillus* and *Micrococcus* and also diphtheroid bacilli. Species that have been isolated from HVAC systems and other parts of buildings and can cause problems are *Pseudomonas* spp., especially *P. aeruginosa, Flavobacterium* spp., *Staphylococcus pyogenes, Serratia marscescens,* and *Legionella pneumophila.* More than thirteen serotypes of *L. pneumophila* have now been identified, all of them pathogenic but some causing the less lethal version of the disease known as

Pontiac fever, after the first definitive episode which occurred in that county of Michigan in 1969. The presence of a mixture of even the harmless species of bacteria in the various parts of an HVAC system may be taken as a sign that the conditions for microbial growth are being supplied, and steps should be taken to clean out dirt accumulations and remove standing water.

FUNGI

The group of microorganisms whose cells have a membrane separating the nuclear material from the rest of the cell protoplasm are classified as eucaryotic and include fungi, yeasts, and molds. These are single- or multi-celled or filamentous organisms whose cell walls are well defined structures containing polysaccharides, and sometimes polypeptides and chitin. They reproduce sexually and asexually often, with an abundance of spores that can be carried in air streams. They are ubiquitous in nature and inevitably enter buildings from the outdoors to set up colonies wherever conditions for growth are favorable. They are very resistant, and once they have established themselves in a niche in a building are very difficult to completely eradicate. Most species are saprophytic; many have been found to be the cause of both infections and allergies in office building occupants and are frequently isolated from various parts of buildings and their HVAC systems and associated ducts, chambers, and voids. From the results of about 200 surveys made in various parts of the world of outdoor airborne spores,[4] the same genera, *Cladosporium, Alternaria, Penicillium,* and *Aspergillus,* accounted for the highest mean percentages. A survey of 11 Florida homes[5] showed a high incidence of these same genera in circulating air. From data based on skin reactivity studies, these four genera also constituted those most prevalent in allergic respiratory disease.[4] Approximately 85% of patients found to be allergic to molds will react to one or more of these fungal allergens. Many other fungal species are routinely isolated from the internals of HVAC systems and their relationship to staff allergies is less clear.[6] Fungal species known to cause infections are *Aspergillus* spp., especially *A. niger* and *fumigatus* which classically cause serious lung infections.

PROTOZOANS

The protozoans are generally more advanced in their structure with more differentiation of parts for feeding, locomotion, and reproduction. Their importance in the indoor environment is that they are able to colonize stagnant water in humidifier reservoirs and improperly drained condensate trays. Aerosolization of their cells occurs directly from the contaminated water as it stands or when the humidifier sprays are reactivated; if the contaminated water dries up, either in its holder or on other parts of the system following carryover, the desiccated cells are entrained in the air supply. Potent allergens may be formed, which if inhaled by susceptible people, give rise to a hypersensitivity response resulting in pneumonitis, humidifier fever, asthma, or allergic rhinitis. Additionally, it is now known[7,8] that these protozoans feed on bacteria, including *Legionella* sp., and that the bacteria can survive, sequestered within the cell protected from biocides in the water, or other unfavorable conditions, to multiply and emerge when better conditions return. Normally these tiny animals will not occur on their own in these stagnant waters within the HVAC systems but appear to require the presence of other microorganisms for nutritional purposes. In most case studies where building-associated hypersensitivity has been reported,[9-11] examination of the humidifier water has shown it to resemble a kind of "living organic stew," with representatives of all the different kinds of microorganisms present. The best way to avoid the problems caused by microbes and protozoans in buildings, therefore, may be to prevent any significant build-up of their numbers in the air supply systems, which may become their incubators and disseminators, by keeping such systems clean and free of water.

TYPES OF MICROBIAL PROBLEMS

Fungi and bacteria isolated from heating, ventilating, and air conditioning systems and other parts of buildings can cause problems in susceptible people in two ways: either by causing an infection or by causing an allergic reaction. With an infection the living organism penetrates the body's defenses and actively colonizes tissues, such as conjunctivae or the respiratory tract. The symptoms may range from slight eye, nose, and throat irritation to multiple deaths from acute pneumonia, as happened in Philadelphia in 1976 with what came to be called Legionnaires' Disease. With allergic reactions, susceptible individuals become sensitized to antigenic material and suffer from symptoms that can range from slight "hay fever type" symptoms, as noted in allergic rhinitis, to the much more serious reactions which may occur with allergic asthma, hypersensitivity pneumonitis, and "humidifier fever" cases. This sensitization may be caused by living microbial or protozoan cells, or by fragments of dead cells, or by toxic waste products produced by them. The affected individuals usually show acute symptoms such as malaise, fever, shortness of breath, dizziness, coughing, rhinitis, and muscular aches and pains, which are reduced on leaving the building for several days and then start up again on returning to it.

SOURCES

Microorganisms may gain entry into a building in many ways. Indeed, they may be built into it, for construction dust and debris along with microbes may contaminate HVAC system components and internals from the time of installation. Since microbes and their spores are small enough and light enough to be carried on air currents, the outside air taken into a building may contain large numbers. Usually they do not travel alone in the air as single microbial cells but tend to clump together in groups and, more frequently, become attached to dust and other particles many times larger than themselves. More should, therefore, be trapped in the medium efficiency filters recommended for installation in commercial buildings than if they were travelling as single cells. Even so, a proportion will still pass the filters, and the more there are in the air presented to filters, the higher the numbers that will pass them. The fresh air intake of an HVAC system is an obvious point for any foreign bodies to enter the system, and it should be protected with a grille of fine enough mesh to prevent birds, moths, butterflies and other insects, as well as dead leaves and other plant materials, from being drawn in. If these living things or their remains are allowed to putrefy, the numbers of microbes present may increase many times and may include species more pathogenic than might normally be found.

Apart from outdoor sources, there are also sources of microbial contamination within a building. We ourselves are a prime source of both animate and inanimate pollution, for whenever we move the action of our clothes rubbing against our skin causes the outer, dead layers of skin to be shed. It has been calculated that on average we each shed about seven million particles and cells per minute and that each of these carries with it an average of four microbial cells.[12] This happens all the time to some extent, but it may increase under particularly dry air conditions. Additionally, we are constantly shedding hair, and under warm conditions we perspire. These processes release parts of ourselves into the environment, and these parts are inevitably accompanied by microbes of one type or another from our bodies. One purpose of an HVAC system is to circulate air, and anything as light as these items are also carried in the air. Thus the return air drawn back to an air handling unit in an office building may be heavily loaded with airborne particulate and microbial contamination. If it is a good, well designed system, this air will be diluted with fresh air and efficiently filtered, and its airborne content will be significantly reduced. If not, then a large number of microbes may be passed through the unit's various pieces of equipment, and some may come to rest in the chill coils, condensate tray, or in the fan chamber, or be carried through to be offered again in the air to be breathed by the occupants. A proportion of those microbes that are deposited within the unit will die, but some will survive until growth conditions are favorable. For some this may simply mean the addition of free water which may be readily available from the condensation of humid air as it passes across chill coils; in a very short time, maybe even a few hours, a small number of microbes will have multiplied to a healthy colony actively spreading across a damp condensate tray. In the same way humidifier reservoirs and any other areas of moisture in the system exposed to the circulating air are likely to become contaminated. Particularly prone to this process is internal glass fiber insulation dampened by water carryover from condensate trays, spray humidifiers, or even steam humidifiers. With its indentations and rough surface, internal glass fiber insulation acts as a dust trap, and when damp forms an ideal matrix for microbial growth.

Once one species of microorganism establishes a foothold in an air handling unit, it is not long before others follow because among these lifeforms there occurs a certain amount of parasitism and symbiosis, some living on others, their waste products, or their dead cells. This is particularly true of periodically stagnant bodies of water that can occur in humidifier reservoirs; at least one outbreak of humidifier fever (or Monday sickness) has been shown[9] to be caused by antigenically active cells, or parts of cells, of a species of amoeba, *Naegleria gruberi,* that lives in the water phagocytosing bacteria. The contaminated water is introduced into the supply air by atomization through spray jets, and some water will fall outside the humidifier where it may evaporate. The amoeba will dry up and either die or sporulate. Dust coming off this area will now contain either spores or dried up fragments of the protozoan that are antigenically active and, if breathed in by susceptible individuals in the occupied area, may cause them to become sensitized, or, if they already are sensitized, they may suffer from an allergic response in the form of the fevers already described. Many other species cause reactions in people in this fashion, including the thermophilic actinomycetes, such as *Thermoactinomyces vulgaris,* often isolated from humidifier systems and air washers.[10]

SAMPLING METHODS

A variety of sampling methods have been developed and recommended for establishing the microbial load present on the internal surfaces of buildings and HVAC systems and in the air they are supplying.[13] Rodac (random organism detection and counting) plates containing trypticase soy agar for bacteria and malt extract agar for fungi may conveniently be used for sampling from surfaces; although certain problems may be encountered in some cases because of high numbers of microbes, such cultures are useful in showing which species of bacteria and fungi are present. These plates are normally incubated at 25°C for 48 h for bacteria and up to 7 d for fungi, with enumeration and identification being done by standard microbiological methods. Incubation at 30°C may be useful for the recovery of bacteria if fungal overgrowth occurs at the lower temperature. Counts may be expressed in the numbers of colony forming units per square inch or square centimeter of sampling area. Additional plates may be incubated at 56°C for several days for the isolation of thermophilic bacteria. Swabs moistened with an isotonic solution may also be used to sample from defined surface areas with subsequent inoculation to appropriate culture media and incubation and examination as before.

Bulk samples of water, ideally of at least 1 l, are required from stored water reservoirs, condensate trays, humidifier water, and cooling tower water for the culture of *Legionella* sp. and protozoans where this is required. Sampling for other bacterial species from water may be done using agar slides dipped in the water, or filter slides backed by an absorbent dried pad of culture medium activated by the sample water. After suitable incubation and identification, the results are given as the numbers of viable bacteria per milliliter of water. This is an effective method to obtain routine total bacterial counts from water and is more convenient than the traditional sampling and shipping to the laboratory of water for serial dilution and culture on solid media.

Several instruments are available for airborne microbial sampling. Three of the more reliable in our hands are the Andersen sampler (six stage and two stage), the Casella Slit Sampler, and the Reuter Centrifugal Sampler. Of these, the Centrifugal Sampler is the most portable and gives sufficiently accurate results for routine sampling. Nutrient agar is used for bacterial culture and a suitable medium, such as malt extract agar, for fungi. The numbers of microbes isolated are expressed in colony forming units per cubic meter of air. The manufacturer's instructions as to calibration and sterilization must be carefully followed to ensure accurate results from all sampling equipment. Gravitational or sedimentation methods of sampling for airborne microbes using settle agar plates or similar are not considered to give accurate representations of the airborne microbial load, either qualitatively or quantitatively, and are not recommended.

Whenever indoor airborne microbial counts are made, several outdoor counts should also be made to act as controls.[13] These outdoor counts should be taken well away from known sources but should also be taken adjacent to the outdoor air intake of the building and close to any potential generators of bioaerosols that may affect the indoor air, such as cooling towers and exhaust stacks. The indoor samples should be collected in areas where no complaints have been recorded as well as known problem areas, ideally, at different times during the day and on different days.

For total fungus spore counts, assessments can be made over time with Hirst-type spore traps. Another method is to use the more readily available filter cassette and personal sampling pumps that allow the spores to be washed off, centrifuged, resuspended in a known volume of liquid, and counted in a hemocytometer.

LEGIONELLA SPECIES

The first identified outbreak of Legionnaires' disease occurred at a convention in a Philadelphia hotel in 1976 when, of 221 people infected, 34 died from a fatal pneumonia. The final identification of the causative organism and its route of infection were not made until many months after the event. Many other cases of *Legionella* have now been identified using immunoserological methods to pinpoint the causative agent, and some sources report that the Communicable Disease Center at Atlanta estimates that more than 40,000 cases occur in the U.S. every year. The *Legionella* bacilli occur naturally in soil and water and may contaminate untreated water used in cooling towers or other water systems where large volumes are stored. For their survival and multiplication, *Legionella* require certain substances to be present in the water, such as iron and an amino acid, cysteine, that may be supplied by other microbes. People have been known to become infected by breathing contaminated aerosols from cooling tower spray drift, from seldom used shower heads in institutions, and from contaminated HVAC systems.

Until recently, diagnosis of *Legionella* contaminated water samples was usually done using culture methods that may take from 4 to 14 days to give results. Now the Legionella Rapid Assay (LRA)[14] test is available which gives a result within one working day from receipt in the laboratory of the water sample. The test is specific for *L. pneumophila* serogroup one, subtype Pontiac, which is responsible for more than 90% of outbreaks, and it can detect as few as 1000 organisms. The assay is a colorimetric monoclonal antibody method based on the well tried and tested Enzyme Linked Immunosorbant Assay (ELISA) principle using antibodies specific to the virulent Pontiac strain of serogroup one conjugated to an enzyme that reacts with a substrate to produce a colored response. The LRA achieves an accuracy comparable with other assay methods presently available, such as immunofluorescence antibody (IFA) techniques, and compared with culture methods it produces more positive results because it can identify the presence of viable but nonculturable *Legionella*. The LRA test is relatively easy to carry out, and a batch of five or ten samples can easily be analyzed in a modestly equipped laboratory within one working day. Test water samples are concentrated, either by filtration or centrifugation, and then inactivated by exposure to 80°C for 10 min. The various solutions and conjugates are prepared and the vacuum manifold set up. The bacteria are then filtered through the sample filter units where they are focussed into a central spot. The filter is then incubated in a solution of monoclonal antibody specific for *L. pneumophila* SG1 (subtype Pontiac) which has been conjugated with the enzyme alkaline phosphatase. Unbound conjugated antibody is washed away and a substrate that turns blue and precipitates in the presence of alkaline phosphatase is then added. The color intensity of the central blue spot indicates the concentration of *L. pneumophila* SG1 (subtype Pontiac) present in the sample. The results from the first 100 samples examined in the HBI Laboratories using the Micro-C LRA test gave 47 positives, of which 18 were from 41 samples of cooling tower water.

INTERPRETATION OF MICROBIAL RESULTS

The correct interpretation of microbial sampling results from the indoor environment is as important as the sampling techniques themselves. Surface culture samples from the internals of air handling units can give rise to misleading results and they must be interpreted with great care. Many variables exist in the symbiotic and pathogenic interrelationships that exist between man and microbes, not least of which is the number of infective or allergenic units necessary to produce symptoms. Other variables include susceptibility of the potential host, which can be related to such factors as age and general health; the local environment, which can include type of occupation and population density; and the mode of operation of the air supply system, which can be affected by changes in ventilation rates and the quality of the hygienic maintenance program. Such surface culture results can be very useful in a qualitative sense in demonstrating which groups of microbes are present and may in this way forewarn of potential problems. For example, if large numbers of the common, dust associated *Bacillus* sp. are identified in an air handling unit, this may not cause as much concern as a finding of only small numbers of fungal colonies identified as *Cladosporium* sp. or *Aspergillus* sp. because of the well-known propensity of the latter to cause allergic reactions and infections in susceptible people. Potential reservoirs and disseminators of microbes may be identified in this way, and comparison with the identification results from airborne samples may confirm such sources.

National standards for levels of airborne microbes in commercial buildings are not yet available, but many "safe levels" ranging over a wide scale have been proposed over the years. Experience of many building survey results[6] suggests that an acceptable level might be 750 cfu m^3 with the provision that if the total airborne count is lower, but among the species isolated are some known to cause infection or allergies in susceptible people, then even that low count is unsatisfactory and steps should be taken to reduce or eradicate the microbes. Such potentially problematical bacterial species include *Pseudomonas aeruginosa* and *Flavobacterium* sp. and among the fungi members of the genera *Alternaria, Cladosporium, Aspergillus,* and *Penicillium.*

In the more visible areas of office buildings, accumulations of dust and dirt are not normally allowed to gather because they are unsightly and unhygienic. The same attitude should be applied to areas of the building that are directly involved with the conditioning and delivery of air to be breathed by the occupants for there is no doubt that in buildings with clean and well cared for HVAC systems with effective air filtration systems and adequate outside air makeup, complaints and problems are significantly less.

In the case of the identification of *Legionella pneumophila* in cooling towers or water associated with air handling units, or in any other situation where a potential for exposure is present, the extent of the risk should be assessed. This should be based on the factors known to enhance the growth rate of *Legionella,* such as optimum temperature, pH, availability of nutritional requirements, water change rate, and the presence or absence of a water treatment program; the feasibility of airborne transmission related to exposure time and droplet size; and the total microbial population in the water, the presence of slime or sludge, and the total legionella count. Each factor is given a numerical rating, and these are compounded to give an overall score — the higher the score the greater the risk. This can be considered only as a guide, for the mere presence of *L. pneumophila* at any concentration does not imply that exposed people will succumb to infection.

However, if the results indicate that high risk levels are present, then it would be considered prudent to deal with the water promptly, ensuring that conditions are not allowed to remain that will allow bacterial or slime growth to occur.

CONCLUSIONS

The study of sick building syndrome is ongoing, and as methods continue to be developed, more will be learned about the true role of microbes and their products in the problem. For example, investigations are being done[15] into the role played by bacterial endotoxins and fungal glucans in SBS for it is known that these substances do cause inflammation and affect the immune system. Dose-response relationships have been established for symptoms produced in occupational environments,[16] but since the amounts of these substances required to produce responses in susceptible people are very small, it could be that small numbers of the appropriate organisms have so far been overlooked or discounted in building studies. Consideration has been given to the role of volatile organic compounds (VOCs) in problem buildings, and it is known that some of the odors produced by fungi are caused by products that belong in this group[17] but it is not certain what part they play in symptom production, either alone or in conjunction with other similar substances. It may be that there is an enhancing of reaction when certain compounds are grouped together. Over the years, some species of bacteria and fungi have become accepted as being "nonpathogenic" or harmless. It may be that this definition will have to be revised if new evidence comes to light showing that a slightly different method of attack has been developed by such species. Similarly, it may be that "new" species of microbes will be discovered that play a part in SBS in the same way that *L. pneumophila* has emerged. In the meantime, it is necessary to continue sampling for microbes in problem and nonproblem buildings alike, from their surfaces, water, and air, so that information can be gathered to help increase our knowledge of their role in SBS.

REFERENCES

1. World Health Organization, Regional Office for Europe, "Indoor Air Pollutants: Exposure and Health Aspects," Report on WHO Meeting, Norlingen, June 8 to 11, 1982, ISBN 92 890 12447.
2. Sykes, J.M., "Sick Building Syndrome: A Review," UK Health & Safety Executive, Technology Division, Specialist Inspector Rep. No. 10, (June 1988).
3. Robertson, J.G., Hearings before the Sub-Committee on Health & Environment of the Comm. on Energy & Commerce of the U.S. House of Representatives, 99th Congress of the U.S.A., 2nd Session on HR 4488 and HR 4546, June 1986.
4. Col, G.T. and Samson, R.A., in *Mold Allergy,* Al-Doory, Y. and Domson, J.F. (Eds.), (Philadelphia: Lea and Febiges, 1984).

5. Binnie, P.W.H., "Airborne Microbial Flora in Florida Homes," Proc. 4th International Conference on Indoor Air Quality, (Berlin, West Germany, August 1987).
6. Healthy Buildings International, Inc., Fairfax, VA., unpublished data, (1981 to 1989).
7. King, C.H., Shotts, E.B., Wooley, R.E., and Porter, K.G., "Survival of coliforms and bacterial pathogens within protozoa during chlorination," *Appl. Environ. Microbiol.,* 54(12):3023–3033 (1988).
8. Richmond, C., "Why Legionnaires' Bacteria are Hardy," *New Scientist,* p28 (December 1985).
9. Edwards, J.H., Griffiths, A.J., and Mullins, J., "Protozoa as Sources of Antigen in Humidifier Fever," *Nature (London),* 264:438–439 (1976).
10. Banaszak, E.F., Thiede, W.H., and Fink, J.N., "Hypersensitivity Pneumonitis Due to Contamination of an Air Conditioner," *N. Engl. J. Med.,* 283:271–276 (1970).
11. Ashton, I., Axford, A.T., Bevan, C., and Lotes, J.E., "Lung Infection of Office Workers Exposed to Humidifier Fever Antigen," *Br. J. Ind. Med.,* 38:34 (1981).
12. Spendlove, J.C. and Fannin, K.F., "Source, Significance and Control of Indoor Microbial Aerosols: Human Health Aspects," *Public Health Rep.,* 98(3):229–224 (1983).
13. Am. Conf. Gov. Ind. Hyg., Comm. on Bioaerosols, "Guidelines for assessment & Sampling of Saprophytic Bioaerosols in the Indoor Environment," *Appl. Ind. Hyg.,* 2(5):R10–R15 (September 1987).
14. The Boots Company, plc, Nottingham, England, "The Micro-C Legionella Rapid Assay User Guide," (1988).
15. Rylander, R., Sorensen, S., Goto, H., Yuasa, K., and Tanaka, S., "The Importance of Endotoxin and Glucan for Symptoms in Sick Buildings," unpublished report, (1989).
16. Palchak, R.B., Cohen, R., Ainslie, M., and Lax Hoerner, C., "Airborne endotoxin associated with industrial-scale production of protein products in gram-negative bacteria," *Am. Ind. Hyg. Assoc. J.,* 49(8):420–421 (1988).
17. Canadian Dept of Health & Welfare, Working Group on Fungi and Indoor Air, "Significance of fungi in Indoor Air," *Can. J. Public Health* 78:S1–13 (March 1987).

CHAPTER 4

Project Designs for the Abatement of Microbial Contamination

E.N. Light*, A.C. Bennett*, D.T. Dyjack*, A. Cooper, K.L. Long, S.R. Lamb, and C.S. Yang

INTRODUCTION

The significance of fungi and bacteria as indoor air quality (IAQ) parameters and their contribution to building-related illness is being increasingly recognized. However, a review of the existing literature indicates that there is little information and even less guidance available regarding decontamination of so-called "moldy" buildings. This chapter briefly reviews some current procedures for managing microbial contamination problems and then focuses on a case study involving assessment and elimination of mold growth in the wall cavities of a school. This is followed by results of monitoring conducted during and after remediation. Discussion includes general principles that may be applicable to the resolution of other IAQ problems caused by excess moisture in buildings.

Contaminated building surfaces present a source of IAQ problems that may be difficult to locate and even more difficult to resolve. Such contamination may be either chemical (e.g., pesticide misuse) or biological (e.g., excess mold growth). This chapter focuses on the latter.

When a building contamination problem is recognized, a complex assessment process may be necessary to map out affected areas along with primary and secondary sources. Objective identification of sources and exposure pathways is necessary for effective decontamination. Poorly planned abatement efforts may be ineffective, at best, or actually serve to increase occupant exposure.[1] Where

*Note: Work completed while primary authors were with Biospherics, Inc., Beltsville, Maryland.

Table 1. Simplified Microbial Assessment

Air Sampling Results	Visual or Historical Indicators of Sanitation Problems	Action?
Normal organisms/ background counts	No sanitation problems	No
Normal organisms/ background counts	Sanitation problems present	Yes
Opportunistic pathogens and/or elevated counts	Indicators of sanitation problems generally present	Yes

microbial contamination is an issue, an understanding of the factors that promote and prohibit growth is essential to remediation.

Excessive mold and/or bacterial concentrations in indoor air primarily impact the health of allergy-prone (atopic) individuals. Such persons may experience the relatively common symptoms of allergic rhinitis or asthma shortly after initial exposure.[2] Much less frequently, airborne microorganisms cause susceptible individuals to develop severe hypersensitivity illness or opportunistic infections.[1,3]

Microbial contamination in buildings can usually be traced back to either unsanitary mechanical equipment (e.g., growth in condensate pans) or excess moisture (e.g., flooding, leaks, condensation, and elevated humidity).[1,4] The extent of microbial growth depends on the type and amount of wet materials, duration of the moisture source, building ventilation, and the timing and type of cleanup. Where major building areas retain excess moisture for extended periods of time without proper disinfection, high hold and/or bacterial concentrations may be expected.

Initial assessment procedures for microbial contamination generally include inspection of the facility, interviews with occupants and maintenance personnel, and, if warranted, air sampling.[5] Air sample results are generally reported in terms of numbers and type of viable organisms. There is no concensus approach to the interpretation of airborne microbial samples. Various assessment schemes have been suggested including specific action levels[3] and consideration of indoor/ outdoor relationships.[6] The authors assign microbial assessment results to one of three categories for the purpose of determining whether a microbial problem exists (see Table 1).

In the first category, only common, nonpathogenic organisms are present and their numbers are within the background range. In well maintained buildings, this is generally under 500 CFU/m^3 (total fungi or total bacteria). In this case, there is no history of unresolved sanitation problems and a visual/odor inspection does not reveal significant indications of unsanitary conditions. No action is recommended.

In the second category, air sample results are also low and unremarkable, appearing to be negative. However, the building history or inspection indicates the potential for significant sanitation problems. Remedial measures are recommended here as a precaution. It is likely in such buildings that higher bioaerosol concentrations will occur at some point in the future if sanitation problems are not addressed.

The third category is where air sampling reveals counts above 500 CFU/m^3 or the presence of opportunistic pathogens. Opportunistic pathogens (e.g., *Aspergillus niger)* can cause infection in persons with weakened immune systems.[7] Visual or historical indicators of sanitation problems are generally also present, but may not always be obvious. Some type of action, ranging from major abatement to further study, is recommended in such cases. The specific response selected is generally based on answers to the following questions:

- How high are microbial counts relative to normal background?
- How dominant are any opportunistic pathogens which are present?
- Are air sampling results consistent?
- What is the likelihood that elevated readings are due to an ongoing sanitation problem?
- Are building-related allergies actually being reported?
- Are immuno-compromised individuals likely to be present (e.g., patients with AIDS or undergoing cancer therapy)?

More detailed characterization of airborne microorganisms may also play an important role in developing abatement strategy. For example, the presence of some microorganisms may serve as environmental source indicators (e.g., high concentrations of *Sporobolomyces* sp. suggest excess moisture). Other fungi present in high levels might indicate possible decay in wood structures.

Consideration of microbial abatement in the general literature focuses on replacement of contaminated porous materials and cleaning/disinfection of hard surfaces.[1,4] In proposed IAQ regulations governing state facilities in New Jersey, the following nonbinding recommendations are offered for microbial decontamination: workers wear respirators; materials removed carefully to minimize aerosolization; surfaces cleaned with HEPA vacuum (high efficiency removal of small particles such as spores) followed by dilute bleach or other biocide.[8] Beyond these basic corrective measures, there is little specific guidance on strategies for conducting major remediation projects. Success of such projects should be dependent on the following factors: (1) controlling occupant exposure during remedial work; (2) systematically sanitizing and protecting surfaces; and (3) demonstrating that microbial air quality has stabilized in the background range.

The following case study involves a school where moisture problems (leaks and condensation) developed in a four room modular addition. Wall and ceiling leaks were noted in conjunction with heavy rains. After several months, stained ceiling tiles were changed and outside walls were waterproofed. Each room (A, B, C, and D) was virtually identical and ventilated by a window fan-coil unit (no central HVAC). The authors were retained as air quality consultants after building-related illnesses were suspected. Data included in this paper was obtained over a 4-month period covering these events:

- Initial assessment (following corrective measures noted above)
- Second assessment (following disinfection of exposed surfaces)
- Preparation of abatement specifications
- Abatement monitoring (demolition of contaminated wall cavities)
- Clearance monitoring

METHODOLOGY

Air sampling was conducted following the general guidelines of the American Conference of Governmental Industrial Hygienists.[5] An N-6 stage Anderson impactor was used to collect 60-s air samples. Standard Methods Agar was used to culture bacteria while Sabouraud Dextrose Agar was used for fungi. Plates were incubated 3 to 5 d (25°C fungi; 35°C bacteria), followed by colony counts. When requested, organism types were also identified.

During abatement, containment pressurization was observed qualitatively with smoke tubes. Leakage of demolition dust from the containment was estimated with the use of a light scattering respirable dust monitor (ppm, inc.).

RESULTS

Assessment

The authors' initial assessment of the complaint area consisted of an inspection, interviews with school personnel, and limited air sampling.

For the 4 months preceding the original on-site study, there had been complaints of musty odor and aggravation of preexisting allergies. Airborne microbial levels were sampled in all four rooms. Fungal concentrations were consistently elevated (up to 1800 CFU/m^3), while airborne bacteria remained in the background range. All fungal samples were dominated by the mold *Cladosporium* sp., a common allergen not of major concern in regard to infections. The observed mold concentrations could be responsible for ongoing allergy complaints expressed by certain teachers in this one section of the school. Visual inspection was inconclusive, with obvious water leaks having been eliminated and no growth or stains observed on exposed surfaces. At this point, school personnel HEPA vacuumed and disinfected exposed surfaces as a preliminary measure while a more detailed response was under consideration.

The rooms were retested 4 d after cleaning. Airborne fungal levels were substantially reduced (maximum 600 CFU/m^3) and bacteria remained in the background range. A few days later, however, musty odors and allergy complaints reappeared, suggesting that the primary source had not been addressed. At this time, slight blistering of the walls was noted in a few locations and it was hypothesized that moisture inside the wall cavities was the primary source of microbial contamination. This was confirmed when a small hole was opened to the wall cavity. Visible growth and strong musty odors were readily apparent. Bulk samples of insulation and wallboard confirmed this to be a significant source with bacterial levels in the 10^5 CFU/gm range (fungal readings were generally low). Further investigation revealed that an improperly installed ceiling vapor barrier had been allowing moisture to condense and drain into the wall cavities. The ongoing nature of the problem was confirmed 1 month later when the rooms were resampled and the peak airborne fungal count had risen about 50% to 950 CFU/m^3.

Abatement Specifications

The facility owner's representatives decided at this time to conduct a detailed abatement project with most of the school to remain occupied during the work. Specific goals were to be as follows:

1. Decontaminate wall cavities
2. Contain emissions during demolition
3. Protect cleaned areas from recontamination
4. Demonstrate that project air quality criteria have been met

With slight variations, precautions commonly used for asbestos removal provided a model for the project.[9] Basic demolition consisted of removing porous, contaminated material from the wall cavities including all gypsum board and insulation. Work areas were cleaned with a HEPA vacuum cleaner followed by disinfection. All work was to be conducted inside a plastic containment under negative pressure. Negative pressurization was achieved with an exhaust fan mounted in an exterior window. Unlike its asbestos counterpart, this exhaust was not filtered. Nonpathogenic spores discharged into outdoor environment were not considered to represent a health hazard.

Although there are several categories of chemicals that can be used as disinfectants.[10,11] Bleach was selected because it was effective against the target organisms, relatively nontoxic, and readily available. Dilutions of 1:5 or 1:10 bleach are commonly used in hospitals, laboratories, etc., for disinfection. Although bleach may be irritating if improperly used, other disinfectants may have more serious side effects (e.g., formaldehyde) or less effective (ammonium-containing disinfectants may actually encourage future microbial growth by leaving residual ammonium as a nutrient source).

Although school was in session during the project, all demolition work was conducted after business hours and buffer zones were designated around active work areas. Work progressed one classroom at a time, with a double layer of heavy plastic sheeting enclosing the demolition area (rear third of each room). Adjacent rooms were kept vacant.

There are no worker protection standards for exposure to nonpathogenic molds and bacteria. The contractor was encouraged to screen out workers with a history of allergies and to promote the use of air purifying respirators with HEPA filters. Disposable coveralls were also optional. Access and egress were through the outside window, making workers' clothing an unlikely route of contamination to unprotected portions of the school.

Following removal of all gypsum board and insulation from the rear wall, the following sequence was followed:

1. Bag gross debris
2. HEPA vacuum surfaces
3. Wipe down surfaces (after at least 1 h settling time) with 10% bleach solution, allowing 20 min contact time
4. Spray cracks and crevices with bleach solution
5. Encapsulate wall cavity (after bleach has dried) with a sealant containing an antimildew agent

These steps were intended to remove materials with excess microbial growth from the wall cavities, effectively kill any remaining organisms, and leave a residual biocide to maintain sanitary conditions. Meanwhile, the exhaust fan would be flushing airborne organisms from the containment. The authors suggested at least 24 h of flushing before clearance air testing in order to provide for a sufficient exchange of air following peak concentrations generated by demolition.

All major work activities were monitored by an on-site industrial hygienist. The containment was inspected regularly, with work to be stopped if barriers were breached or musty odor, haze, or excessive dust appeared on the outside. Microbial air sampling was conducted daily when work was ongoing. A respirable particulate monitor was used on some occasions to evaluate the effectiveness of the containment as a dust barrier. Any elevated readings outside the containment were to result in more stringent precautions and disinfection of affected areas.

A formal clearance procedure was established to be repeated in each classroom. First, cleanup had to be completed to the visual satisfaction of the industrial hygienist. Second, air samples for bacteria and fungi collected in the containment had to be either less than 500 CFU/m^3 or lower than outside air concentrations (whichever standard is greater). Third, the entire classroom must pass a similar clearance test after the containment is removed and surrounding area cleaned. Detailed specifications for abatement can be found in Appendix A.

The building owner selected a general construction contractor to implement these specifications. Although a contractor more experienced in the handling of hazardous materials and disinfection would have been better suited to the task, the contractor selected had built the original structure which was still under warranty. The project consisted of four separate containments (classrooms A, B, C, and D). Although not intentionally a controlled experiment, conditions differed to some extent between the containments.

First Phase

The initial abatement period lasted 1 week, during which rooms A and B were contained, ceiling vapor barriers in rooms A and B were removed, two containments constructed, interior wall cavities accessed, cleaning and disinfection completed, and walls reconstructed. Major specification enforcement problems occurred due to failure to maintain negative pressure in the containments (improper ventilation) and the reconstruction of room B before it passed a clearance test.

During demolition in room A, fungal levels inside the containment rose to over 10,000 CFU/m³ while bacterial levels were also elevated. Substantial leakage occurred to buffer zones which had counts similar to the containment. The adjacent corridor remained at background levels.

The first containment passed the clearance test (containment level less than outside air) 2 d after it was disinfected. Simultaneous sampling in the room A buffer zonc still showed marginal contamination.

Testing during demolition in room B showed leakage into the outer room area was substantial. Particulate measurements in the outer room were also relatively high. Room B failed clearance 2 d later with fungal counts being higher than outside air. Also, the room B buffer zone remained contaminated. In violation of the specification, the contractor reconstructed the wall in room B while these laboratory results were pending. During the preceding work, the plastic barrier remained in place while all demolition and disinfection tasks were completed. Negative pressure was not maintained because of several deficiencies. These included the exhaust fan being two small and leaking around the edges (short circuiting), makeup air not being drawn through much of the containment, and the fan not running continuously (it was shut down following each work shift). These factors may have all contributed to the leakage noted. Failure to pass the initial clearance test in room B may have been due to an insufficient time between last disturbance and sampling (less than 24 h for air flushing). Incomplete disinfection may have also contributed.

Second Phase

Due to the technical and enforcement problems cited above, abatement activity stopped for 3 weeks. It was determined that the wall in room B would have to be reopened and disinfected a second time. Abatement would then proceed to rooms C and D under continuous negative pressure. A larger fan was to be installed in each containment, sealed into the window frame and operated around the clock from start of demolition to clearance. Demolition was to be permitted only when there was a negative pressure as indicated by the plastic barriers being drawn into the work area. Detailed smoke tube testing was also conducted to verify air movement through the containment.

With heating units off and the onset of winter weather, condensation began to form on surfaces in the work rooms. Space heaters were added to control this new moisture source. Beams that had become wet were disinfected as a precaution.

During the second demolition of the room B wall, containment levels again became elevated. Buffer zone concentrations were somewhat lower than they had been before for fungi but higher for bacteria indicating that some leakage was still taking place. Adjacent corridor readings were low.

Room B passed the clearance test 24 h later, and the buffer zone returned to background levels. However, readings in the outside corridor were now higher.

Custodial sweeping had occurred a few minutes before the air test and may have been the source of this elevated sample.

Abatement of room C proceeded without incident. Inside levels were high during demolition with some leakage to the buffer zone indicated. Corridor levels were also elevated, although this may have again coincided with elevated dust from recent sweeping. The following day, levels returned to background and the containment cleared.

Rooms C and D cleared 1 d after demolition, with the buffer zone and corridor returning to background levels. All four classrooms were retested for final clearance after reconstruction, removal of the containment, and final cleaning. All areas, including the corridor, showed background levels of fungi and bacteria in the final samples.

One possible cause of leakage during the second round of demolition was worker access to the containment. Workers entered through an outside window, which was observed to short-circuit the fan and eliminate the negative pressure. When the access window was opened, significant containment leakage may have occurred into the classroom.

Data Summary

The progression of airborne fungal and bacterial concentrations as they occurred in each classroom over the course of this project is presented in Figures 1 to 4. Fungi levels consistently remained higher than bacteria. Classroom A provides an example of how airborne fungi changed over time. The original assessment (samples during allergy complaints) recorded a level of 1800 CFU/m^3. This temporarily dropped to 250 CFU/m^3 following initial (superficial) cleaning, then rose back to 880 CFU/m^3 a few weeks later. At the height of demolition, the containment exceeded 10,000 CFU/m^3 (maximum limit of detection). The initial clearance level inside the containment following disinfection dropped to 670 CFU/m^3. Final clearance of the room after cleanup was 35 CFU/m^3 (see Figure 1).

Containment leakage occurring during the project is illustrated in Figures 5 through 7. As mentioned previously, background counts under sanitary building conditions can generally be expected to remain in the background range (below 500 CFU/m^3). During demolition in containment B, fungal counts were very high, exceeding 10,600 CFU/m^3. At the same time, sampling in the buffer zone (unprotected outer portion of classroom C) reached 5300 CFU/m^3. This is elevated for an interior environment with the only apparent source being containment leakage. After 1 to 2 d, all buffer zones where such contamination was documented had returned to the background range (350 CFU/m^3). Sampling results from the corridor (adjacent to the upper zones) are summarized in Table 2. Readings generally remained low. Elevated concentrations appeared to be related to local custodial activity rather than the abatement.

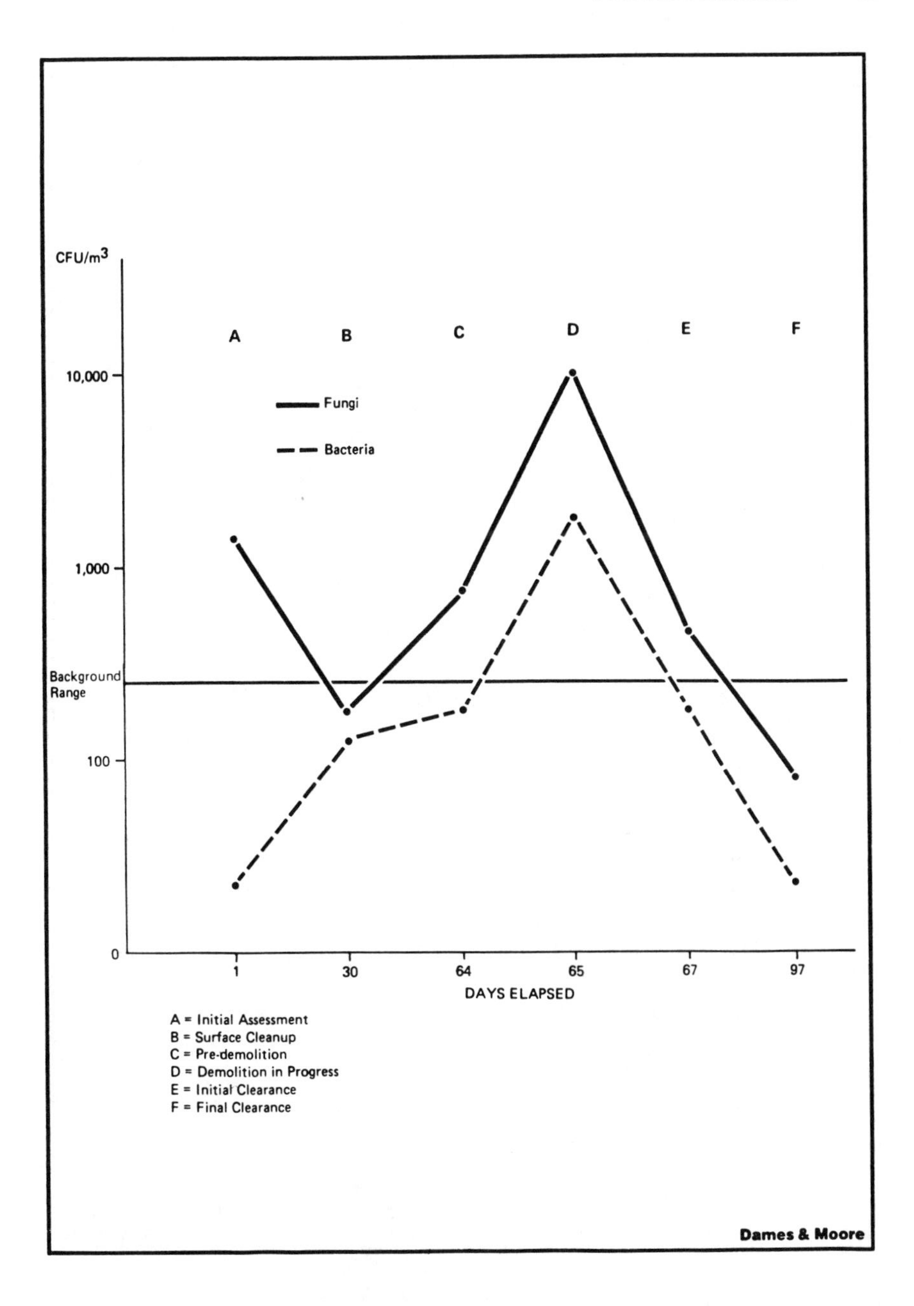

Figure 1. Bioaerosol concentrations in room A.

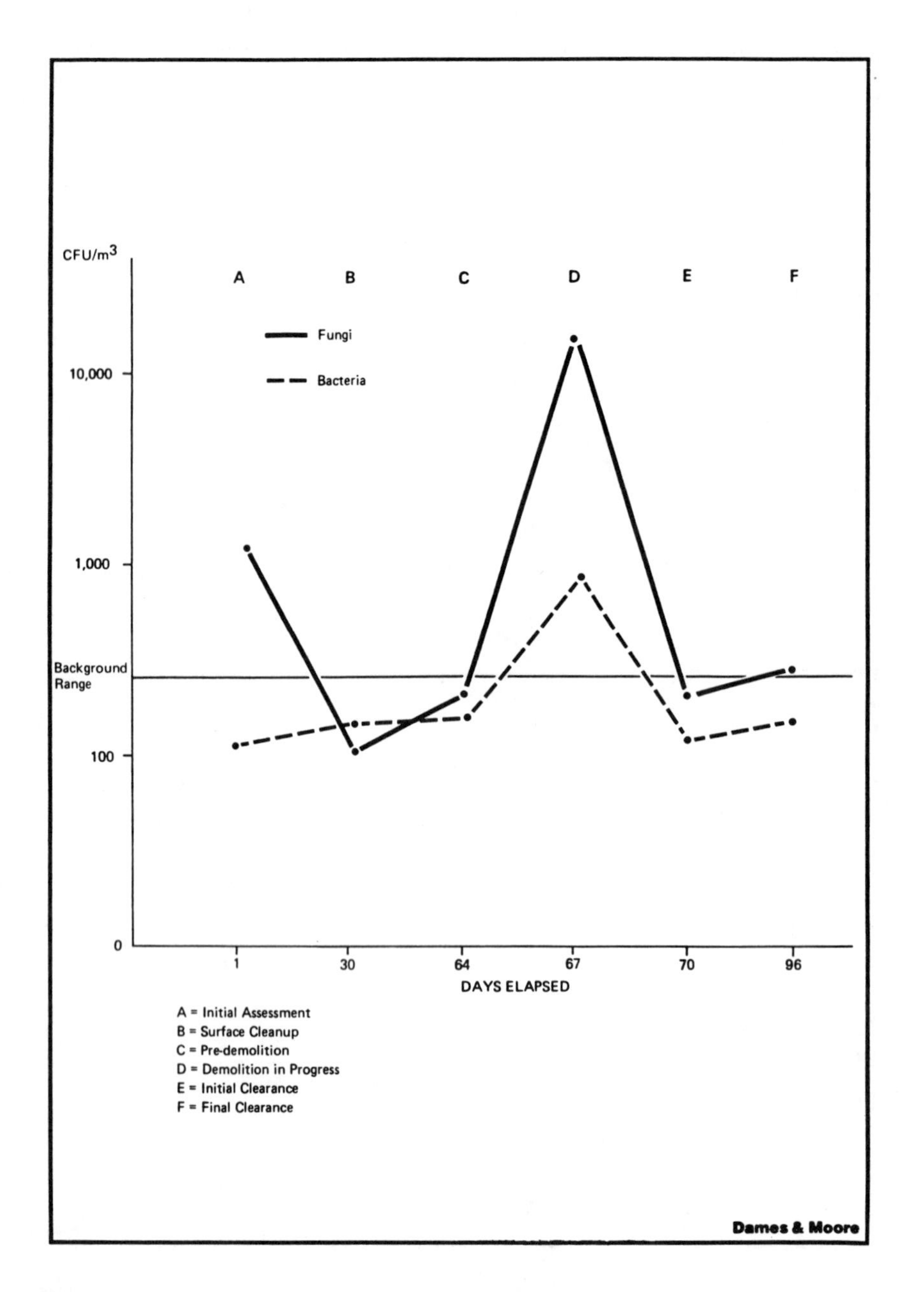

Figure 2. Bioaerosol concentrations in room B.

DISCUSSION

Air monitoring was utilized during the abatement process to provide general documentation of project conditions, identify leakage from the containment, and

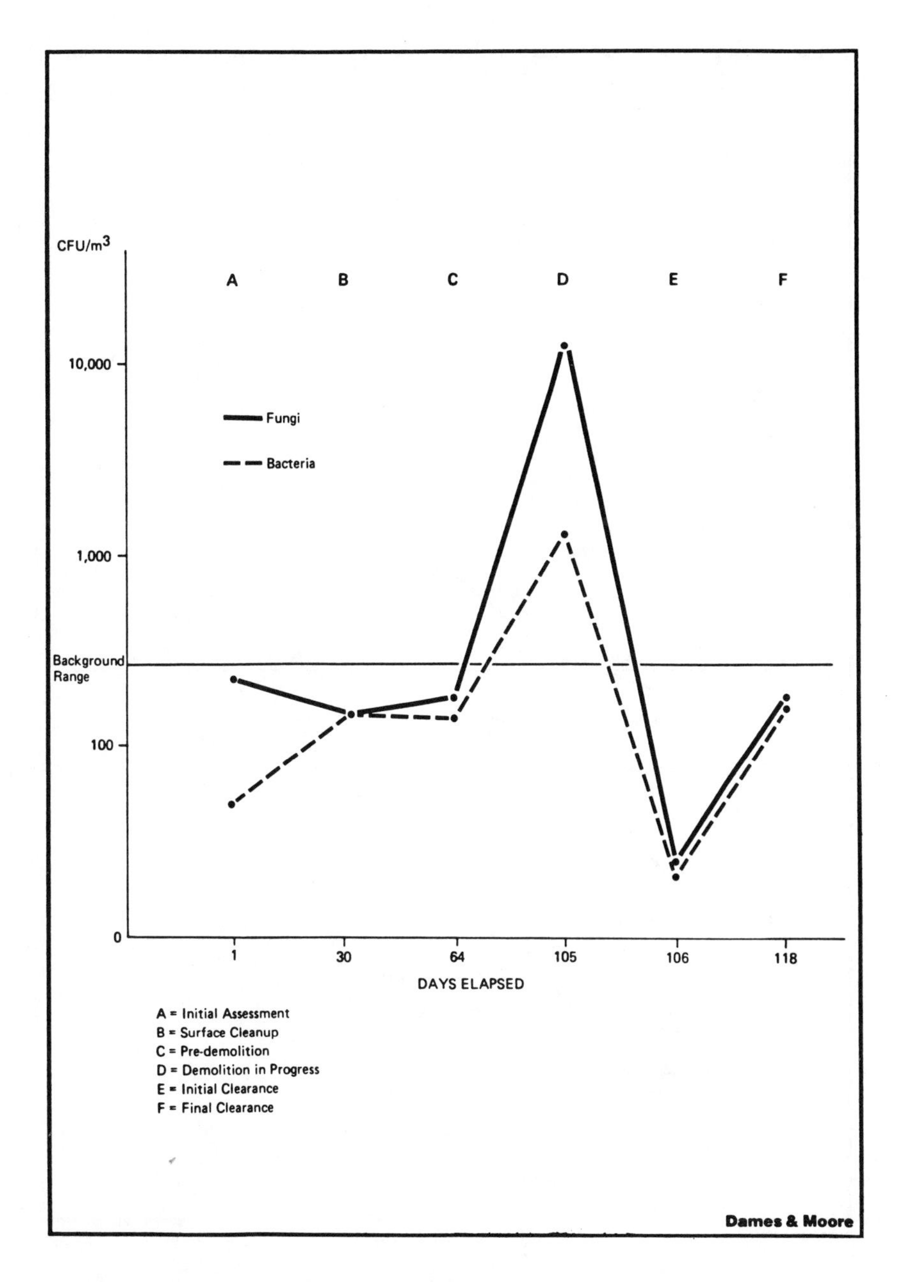

Figure 3. Bioaerosol concentrations in room C.

determine when clearance goals had been achieved. Microbial air sampling required a minimum of 3 d incubation time in order to make a preliminary count of viable organisms. This delay detracted from its utility in terms of timely correction of containment leaks on a timely basis and proved to be disruptive to project scheduling.

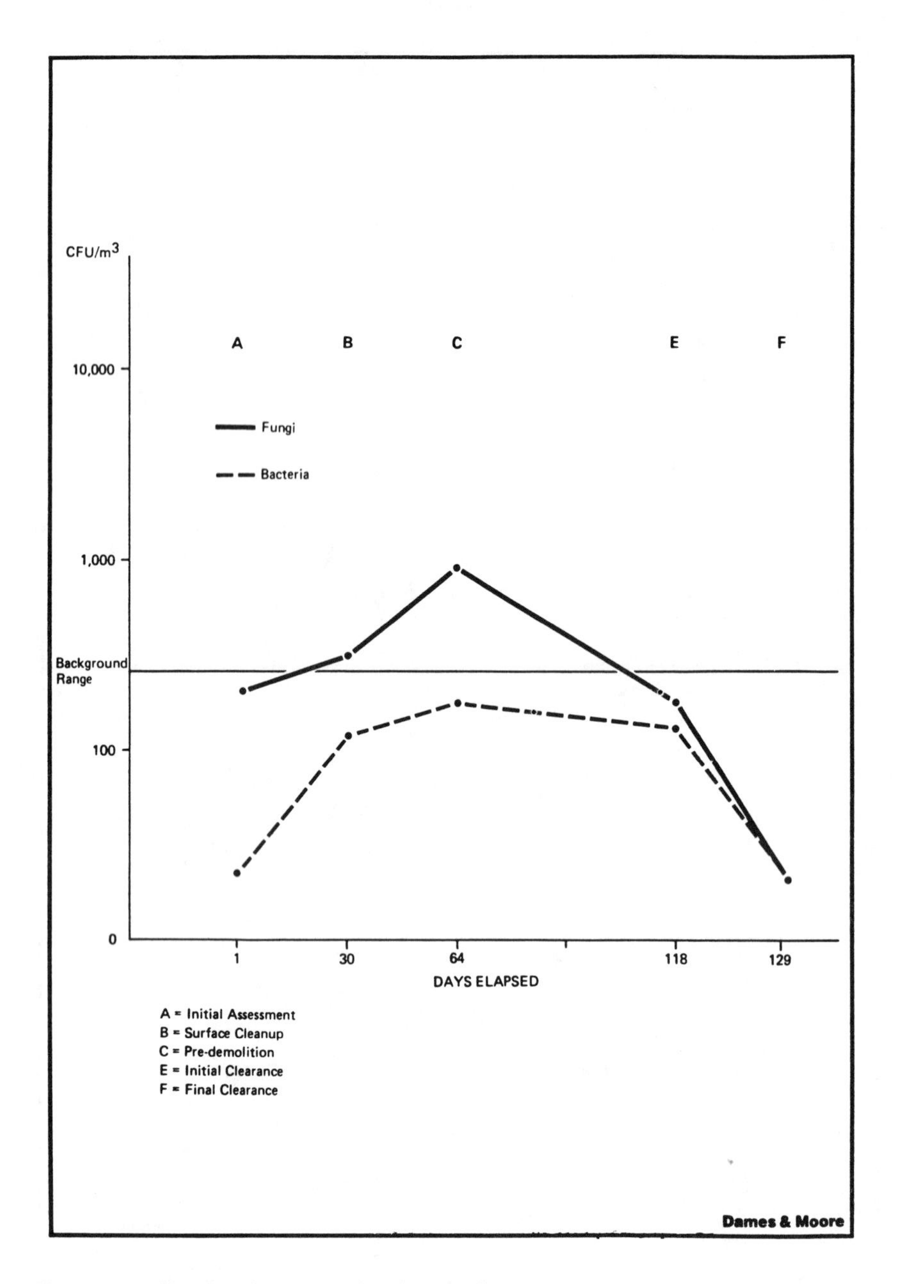

Figure 4. Bioaerosol concentrations in room D.

On-site inspector observations did provide some helpful, real-time feedback in regard to containment integrity. On one demolition day, a direct reading aerosol monitor was used to monitor the containment. Although only limited data was obtained, readings were elevated where high microbial counts were later obtained. This instrument indicates a background range for respirable dust in most buildings

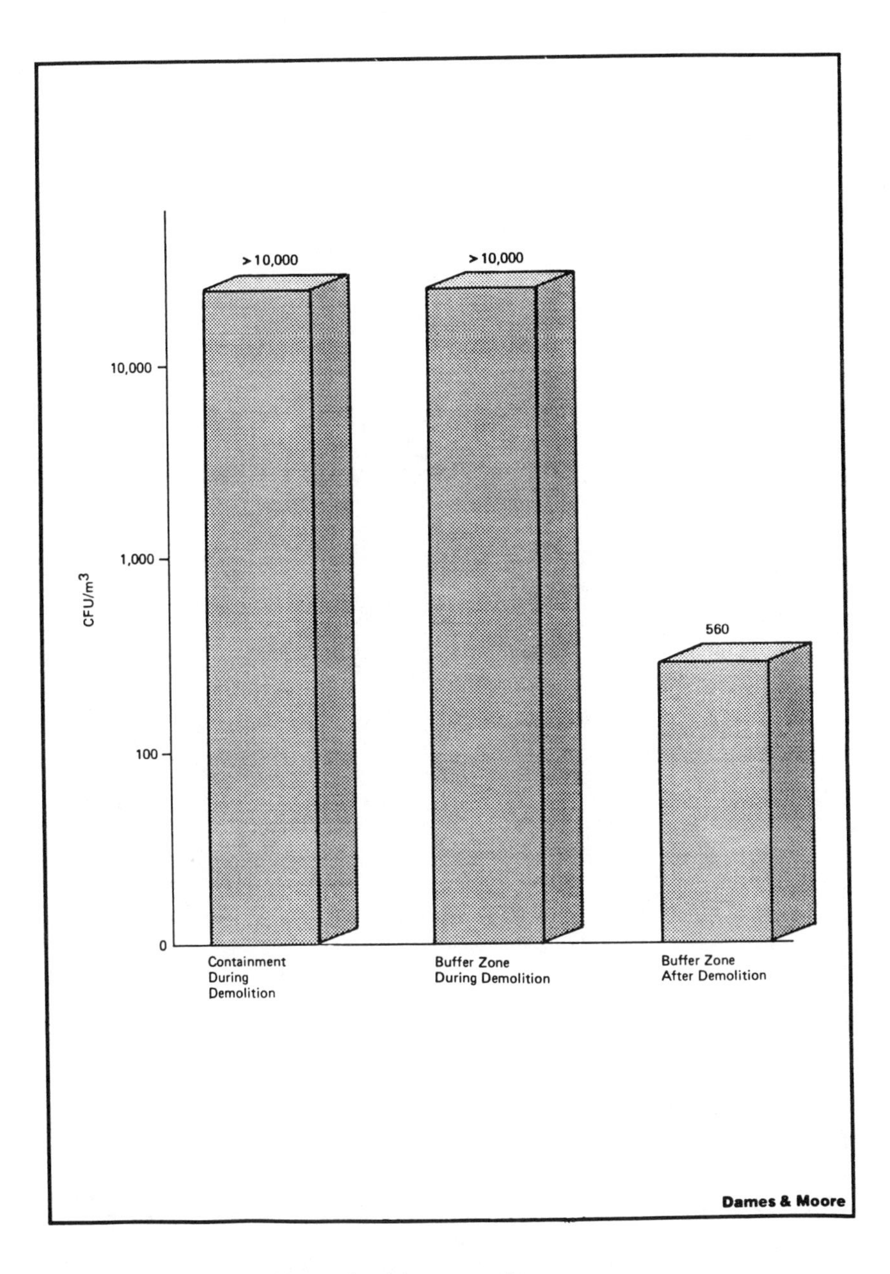

Figure 5. Leakage of airborne fungi from room A.

without significant particulate sources of 5 to 50 μg/m³. On the day of demolition in room B, while background levels of respirable suspended particulates (RSP) in the school were around 10 μg/m³, levels outside the containment in the room B buffer zone had risen to 300 μg/m³. Airborne microbial data collected from the same location had a fungal count exceeding 10,000 CFU/m³.

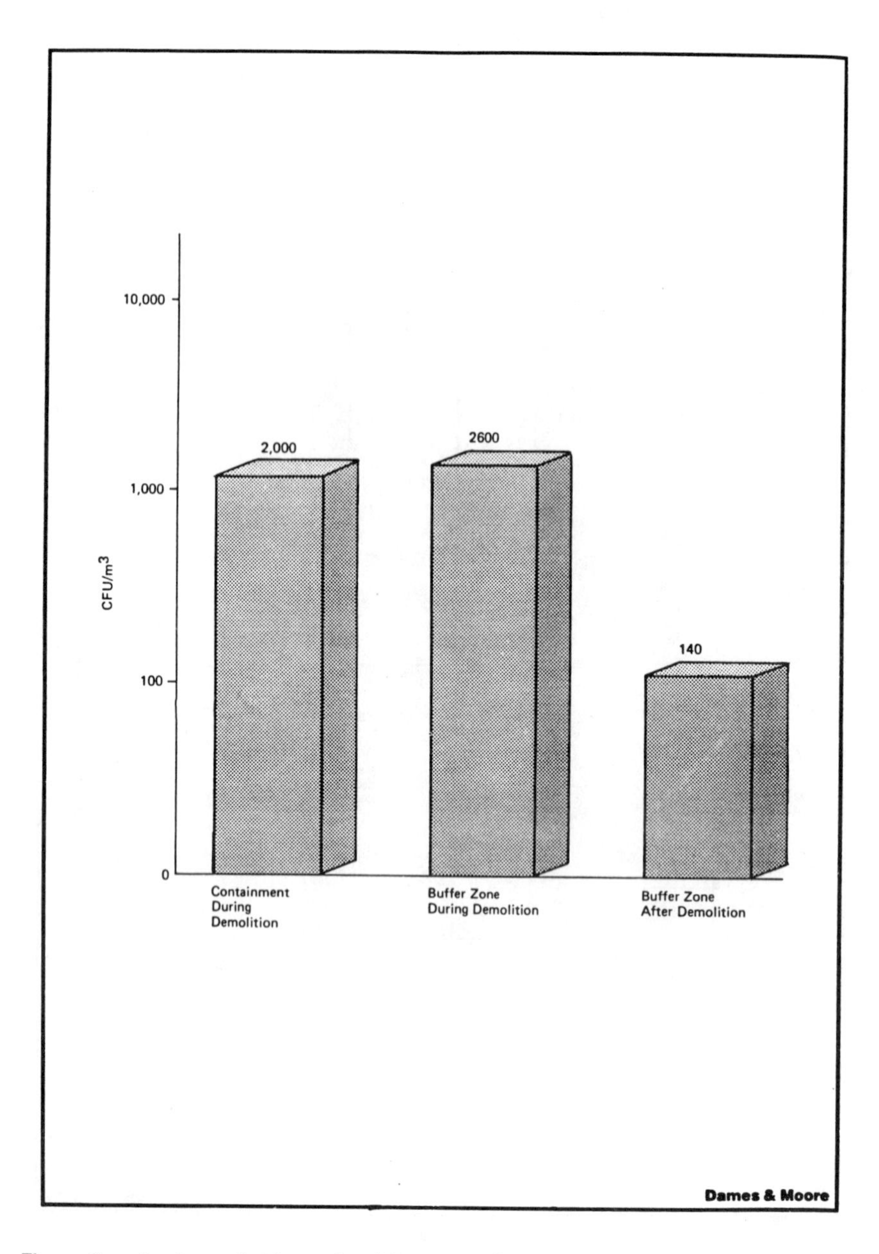

Figure 6. Leakage of airborne fungi from room B.

Further work may show the aerosol monitor based on light scattering to be a useful tool where microbial abatement involves the generation of dust (such as during demolition). Although individual bacteria and mold spores are often too small to be recorded effectively by such instrumentation, bioaerosols correlate under some conditions with general dust levels, especially when substantial dis-

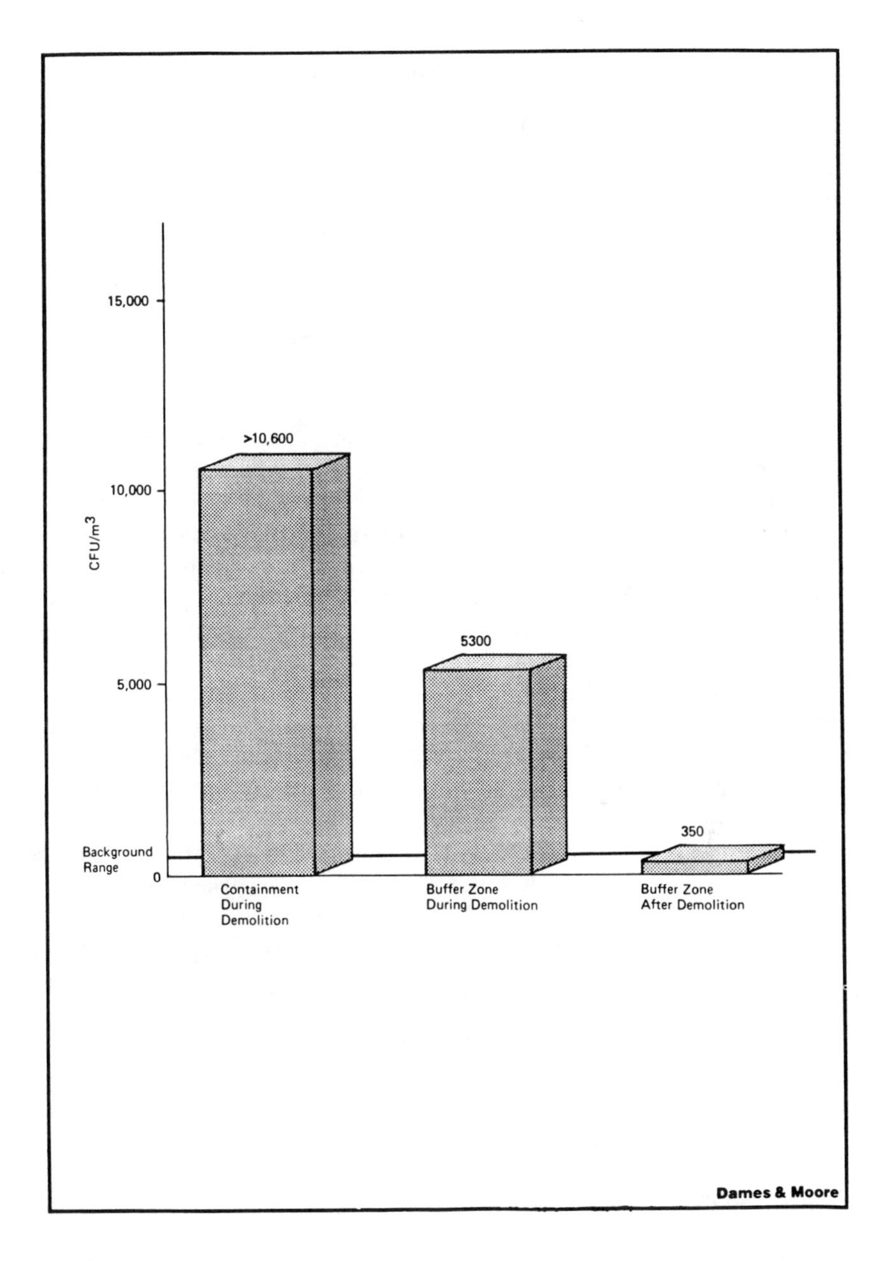

Figure 7. Leakage of airborne fungi from room C.

turbance of contaminated materials is ongoing. This general indicator can be available immediately to those regulating the abatement process while more specific microbial sampling results are being developed.

Another possible approach to the problem of sample turnaround time during microbial abatement involves spore trapping. One method being investigated by

Table 2. Bioaerosol Concentrations in Adjacent Corridor

Project Day	Bacteria CFU/m³	Fungi CFU/m³	Project Status	Other Activity
37	110	<35	Day after demolition	—
94	280	<35	During demolition	—
95	1500	780	Day after demolition	Sweeping
104	71	3600	During demolition	Sweeping
117	460	490	Day after demolition	—
129	250	<35	All work completed	—

the authors involves passing a known volume of air through a cellulose ester filter, which is then cleared and mounted for examination under a phase contrast microscope (similar to procedures for sampling fibers-in-air). Although bacteria are too small for analysis by this technique, fungal spores can be distinguished and counted under the supervision of the mycologist. This may allow a quantitative estimate of total (viable and nonviable) spores within a few hours, bypassing the incubation step. With further development and validation, such a procedure could expedite decision making at microbial abatement projects where airborne fungi are the most critical measurement.

One final issue highlighted by this project verified was importance of tracking outdoor microbial levels. On some days, outdoor fungal counts were very high (e.g., 10,000 CFU/m³) while on other days they were very low. Unfiltered outside air was drawn into containments at various times with an obvious potential to influence both demolition monitoring and clearance results. Under the procedure adopted, when outdoor counts were elevated, containment levels simply had to be lower than outside to achieve clearance. However, on days when outside counts were very low, readings up to 500 CFU/m³ would still be allowed inside in order to be consistent with the "normal" background range. Follow-up testing of each room after the containment passed initial clearance testing indicates that airborne microbial concentrations remained in the background range (see Figures 1 through 4). Monitoring of future microbial abatement projects might be facilitated by locating access points that prevent unfiltered outside makeup air from being drawn into the containment.

CONCLUSIONS

1. Moisture accumulation in wall cavities was the primary source of microbial contamination and IAQ complaints at the school investigated.
2. Stringent control over the microbial abatement process was justified by the sensitivity of the building population (school children and atopic faculty members) and potential spore release during wall demolition. As a result, work was conducted after hours, inside a containment surrounded by unoccupied buffer zones.

3. Levels of airborne fungi and bacteria increased inside the containments during demolition. Leakage into adjacent buffer zones during demolition was apparently caused by loss of negative pressurization within the containment. Contamination of the buffer zone was short-term and was readily eliminated by standard disinfection techniques.
4. Use of a general contractor with no experience in disinfection or the handling of hazardous materials led to enforcement problems resulting in technical errors and major project delays.
5. Air monitoring based on counts of viable organisms resulted in 3- to 5-d delays in the identification of containment leaks and determination of area clearance. Use of a direct-reading aerosol monitor might provide an effective monitoring supplement when remediation generates dust (e.g., demolition). Direct counting of spores collected on cellulose ester membrane filters may also prove to be a useful sampling tool during microbial abatement.
6. A formal clearance procedure was followed to determine when each work area was safe for reoccupancy. This included visual confirmation by an industrial hygienist that the work specifications had been followed. Clearance sampling results in terms of fungal and bacterial counts had to be below 500 CFU/m^3 or, where outdoor counts were higher, less than outside concentrations measured at the same time. This criteria appeared to provide a reasonable basis for concluding that the work area had returned to a stable, background level.
7. Despite several problems encountered, the remedial measures employed in this case study successfully eliminated microbial contamination. These included replacement of gypsum board and insulation in wall cavities, HEPA vacuuming, application of bleach solution, and treatment with a sealant containing an antimildew agent.

RECOMMENDATIONS

1. The extent to which special precautions are needed during the remediation of microbial air quality problems should be based on the degree that contaminated materials must be disturbed and the potential health risk to exposed individuals.
2. Microbial abatement projects perceived as presenting major risks to occupants should be conducted inside contained work areas. Abatement workers should ideally be experienced in both the handling of hazardous materials and disinfection techniques. Such projects should be supervised by an industrial hygienist.
3. Clearance of microbial abatement areas for reoccupancy should be based on verification that specified work practices have been completed and confirmation that air contaminants have returned to background levels.

APPENDIX A

Specification for Microbial Decontamination

SUMMARY

Gypsum board and insulation will be removed from rear (exterior) wall and the wall cavity disinfected. All work shall be performed within an asbestos-type containment to control the release of mold and bacteria. Work shall be conducted after-hours in one-room segments. Work will be stopped and corrective actions taken if monitoring conditions do not meet standards established by this specification. The wall can be reconstructed and barriers removed when clearance criteria are met.

Special equipment/materials needed:

- HEPA filter. A high efficiency particulate air filter capable of removing particles greater than 0.3 mm in diameter with 99.97% efficiency.
- HEPA Vacuum. A vacuum system equipped with HEPA filtration.
- Negative pressure ventilation system. A portable exhaust system capable of maintaining a constant low-velocity air flow into contaminated areas from adjacent uncontaminated areas. For this project, a Microtrap or equivalent shall be used in each work room.
- Disinfectant. Bleach in a 1:10 dilution with water.

SITE PREPARATION

Monitoring

Prior to site preparation, collect background samples for airborne bacteria and fungi (counts only) in each occupied classroom and the outside air.

Adjacent Areas

Room in which containment is to be located shall be vacant for duration of activity. Staff may enter room to collect materials. Fan/coil unit in the work area shall remain off. All interior surfaces shall be cleaned and disinfected and be sealed with plastic so as to prevent any dust from entering during demolition. A plastic drop cloth shall be placed over the floor in the room outside the containment.

Containment

Unless otherwise noted, seal all walls and floors with two layers of six mil plastic sheets. Seal off all openings, doors, windows, fan coil units, light fixtures, etc., with two layers of six mil plastic. Remove ceiling tiles; if there is any water-damaged insulation, remove carefully and disinfect underlying surface. Extend the plastic sheeting from the floor to the roof and wall-to-wall, making a containment of about 3000 ft^2. Ensure that barriers are effectively sealed and taped. Repair damaged barriers and remedy defects immediately and visually inspect enclosures prior to each workday. Use smoke methods to test effectiveness of barriers. Barrier must remain in place until the clearance criteria are met.

Access

Curtained Doorway. A device to allow worker ingress or egress from the outside only through a window while permitting minimal air movement, constructed by placing two overlapping sheets of plastic over a window, securing each along the top of the doorway, securing the vertical edge of one sheet along one vertical side of the doorway and securing the vertical edge of the other sheet along the opposite vertical side of the doorway. Other effective designs are permissible. All waste shall leave the containment through this window.

Negative Air

Negative pressure ventilation units or exhaust fans will work continuously from the start of demolition until clearance is granted. Makeup air and exhaust shall be located so as to provide for the flow of air throughout the containment. The primary source of makeup air shall be from inside the building through a flap-type air lock in the containment barrier.

DEMOLITION

Worker Protection

There are no mandatory respirator standards for work around environmental molds and bacteria. Air purifying respirators equipped with HEPA filters are recommended. Individuals with preexisting allergies should avoid this type of work area, if possible. Dust masks offer some, but not complete protection. Full body disposable protective clothing is optional except in circumstances where a worker must enter the school from the containment. This should be avoided if at all possible and the suite (including foot coverings) removed before leaving the containment.

Removal

Minimize dust generation. Mist with water as needed to suppress dust. Inspect exposed wall cavity to determine moisture sources, document damage, and confirm scope of work with owner. The industrial hygienist must verify that removal of potentially contaminated porous materials is complete.

Cleaning

All surfaces in the containment shall be cleaned in the following sequence:

1. Remove any gross debris.
2. Vacuum clean (HEPA).
3. After a minimum settling time of 1 h, damp wipe with disinfectant (e.g., 10% bleach solution). All surfaces must stay wet for a minimum contact time of 20 min. Cracks and crevices in wall cavity shall be sprayed with 10% bleach.
4. After all surfaces pass a visual inspection by the industrial hygienist, the contractor shall encapsulate the wall cavity and sill plate with sealant containing appropriate biocide approved by the industrial hygienist (e.g., antimildew agent).

Monitoring

Work is in progress

Project will be inspected to ensure it is in conformance with this specification and to conduct periodic air quality tests to document the containment effectiveness. Work must stop and corrections made if there is musty odor, haze, or excessive dust outside the containment, or work is not being performed in conformance with this specification.

Clearance

Final air samples will be collected after all surfaces in the containment are inspected and are found to be dry and free of any debris or settled dust. Aggressive microbial air samples (exhaust fan running) collected in the containment must be less than 500 cfu/m^3 or less than outside air levels collected at the same time.

Reconstruction

Reinsulation and installation of wallboard shall be conducted within the containment after air samples are cleared by the industrial hygienist.

Reestablishment of Classroom

Removal of barriers and reinstallation of systems shall be conducted with a small exhaust fan in an open window. Any necessary repair work shall be completed along with one final cleaning of the room. Filters shall be changed in fan coil unit and ceiling tiles reinstalled.

Final Testing

Air quality shall be retested after the above procedures have been completed in all areas to ensure that major sources of microbial contamination have been controlled.

REFERENCES

1. Burge, H.A., "Approaches to the Control of Indoor Microbial Contamination." *Proceedings of IAQ 87, Practical Control of Indoor Air Problems,* (Atlanta: ASHRAE, Inc., 1987).
2. Health and Welfare Canada Working Group on Fungi and Indoor Air, "Significance of Fungi in Indoor Air: Report of a Working Group," Canadian Public Health Association. *Can. J. Public Health,* 78:2 (1987).
3. Morey, P.R., Hodgson, M.L., et al., "Environmental Studies in Moldy Office Buildings: Biological Agents, Sources and Preventative Measures," *Ann. Am. Conf. Gov. Ind. Hyg.,* 10:21-34 (1984).
4. Morey, P.R., Clere, J.L., et al., "Studies on Sources of Airborne Microorganisms and on Indoor Air Quality in a Large Office Building," *Proceedings of the 1986 ASHRAE Conference on Indoor Air Quality,* (Atlanta, 1986).
5. ACGIH, "Guidelines for Assessment and Sampling of Saprophytic Bioaerosols in the Indoor Environment," *Appl. Ind. Hyg.,* 2:R10–R16 (1987).
6. Morey, P.R., "Microorganisms in Buildings and HVAC Systems: A Summary of 21 Environmental Studies." *Proceedings of IAQ 88, Engineering Solutions to Indoor Air Problems,* (Atlanta: ASHRAE, Inc., 1988).
7. Raper, K.B. and Fennell, D.I., *The Genus Aspergillus,* (Baltimore: Williams & Wilkins, 1965).
8. New Jersey Department of Labor, 1989. Standards for Indoor Air Quality (proposed). *Safety and Health Standards for Public Employees,* (January 23, 1989).
9. U.S. Environmental Protection Agency, "Guidance for Controlling Asbestos-Containing Materials in Buildings," EPA 560/5-86-024, (1985).
10. Joklik, W.K., Willett, H.P., Amos, D.B., and Wilfert, C.M., *Microbiology,* 19th Ed. (Norwalk, CT: Appleton & Lange, 1988).
11. AIHA Biohazards Committee, *Biohazards Reference Manual,* (American Industrial Hygiene Association, 1985).

CHAPTER 5

Sustained Reduction of Aerobiological Densities in Buildings by Modification of Interior Surfaces with Silane Modified Quaternary Amines

Richard A. Kemper and W. Curtis White

INTRODUCTION

Building-related illness (BRI) or sick building syndrome (SBS) continues to stimulate global attention as the scientific community investigates causative factors and the scope of effects. Many deleterious symptoms, including erythema, mental fatigue, high frequency of airway infections, hoarseness, wheezing, itching and nonspecific hypersensitivity, nausea, headaches, lethargy and dizziness affect the health and productivity of workers.[1]

The onset of these symptoms is insidious and usually attributed to factors other than BRI/SBS. After repeated attacks, however, workers recognize a typical pattern: symptoms appear 1 to 2 h after arriving at work and disappear 3 to 4 h after leaving. These symptoms are the classic manifestations of BRI/SBS. Additionally, workers report that the severity of the attacks usually increases with subsequent exposures. Efforts to determine the etiologic factors, sources of the problem, and effective solutions have proven to be formidable.

BRI/SBS was believed to result from occupant exposure to excessive levels of organic vapors, noxious gases, or physical irritants within closed, tightly sealed buildings. Bioaerosols were identified as causal in fewer than 5% of the outbreaks investigated by NIOSH.[2,3]

The importance of bioaerosols and biogenic materials as indoor environmental pollutants is increasingly recognized. They are implicated as etiologic agents in numerous outbreaks of BRI/SBS and other respiratory illnesses. Inhalation of mycotoxins and aflatoxins has been shown to induce mycotoxicosis (liver cancer) and many are known to be acutely toxic. The long-term exposure hazards to building occupants are not presently known, but currently available data suggest that exposure to mycotoxins could have deleterious effects on health.[4,5]

By design, energy-efficient buildings concentrate the level of airborne microorganisms and their by-products as sourced from environmental surfaces, people, dust, and furnishings, causing them to rise above the threshold at which many occupants will present with a response. Supporting this, Dr. Harriet A. Burge and her research team presented evidence that fungi within tightly sealed buildings can cause hypersensitivity pneumonitis, a condition that may produce permanent lung damage and even death.[6]

Furthermore, researchers at the Walter Reed Army Institute of Research in Washington, D.C., conducted a 4-year study of barracks housing more than 400,000 recruits to examine the incidence of influenza and other respiratory illnesses. The researchers, led by Dr. John F. Brundage, found that trainees housed in modern barracks were about 50% more likely to contract a respiratory infection during the 7-week training period than those housed in older, more drafty buildings.[7]

MICROORGANISMS

Bacteria, fungi, viruses, and algae are all associated with the indoor environment of buildings, and many are capable of producing the symptoms associated with BRI/SBS. Of these, bacteria and fungi are most frequently associated with hyperresponsive illnesses, infections, and toxic response.[8]

Although a building may be infested during construction (particularly with fungi), more typically the organisms are routinely brought into the building by its occupants. Lofted into the air by normal activities in the building, these microorganisms can be transported throughout the building by occupants and the HVAC system. Thus, even the most remote areas of the building become vulnerable to infestation. Under favorable conditions these microorganisms proliferate and colonize interior surfaces.

For example, bacteria play an important role as part of the body's microflora and, along with skin, are shed continuously. Given acceptable growth conditions, they can multiply from one organism to more than one billion organisms in just 18 h. Fungi — typically outdoor organisms known as mold, mildew, and yeasts — enter the building on clothing, are wafted in through open doors, or are pulled in as "makeup air" by the HVAC system.

Inhalation of these microorganisms, their somatic parts, and/or their by-products may produce an immunologic response that triggers the release of specific antibodies. Repeated exposures magnify the antigen-antibody reactions, lowering tolerance levels and exacerbating clinical symptoms. Other manifestations of excessive microbial presence include odors, discoloration, deterioration, and defacement of contaminated surfaces.

ANTIMICROBIALS

Antimicrobial agents have been used for many years to reduce microbial populations and their associated problems. By definition, an antimicrobial agent is an agent that destroys or inhibits the growth of microorganisms. Bacteria, fungi (mold and mildew), yeasts, and algae are the major classes of microorganisms.

Antimicrobials treatments differ in chemical nature, mode of operation, durability, effectiveness, toxicity, safety, and cost. They can be divided into two major categories: bound and unbound. These terms refer to whether or not the antimicrobial has the capability to chemically bond to the surface on which it is applied.

Unbound Antimicrobials

An unbound antimicrobial cannot be bonded to a surface in order to function properly. It must diffuse from the treated substrate and be consumed by the microorganism in order to be effective. Once inside the organism, the chemical agent will act like a poison interrupting some key metabolic or life-sustaining process of the cell, causing it to die. Once the antimicrobial is depleted or is washed away during regular maintenance, protection vanishes. Therefore, the degree of durability desired must be considered when choosing an antimicrobial treatment.

After application, an unbound antimicrobial continues to diffuse or leach from the substrate on which it has been applied. As this diffusion continues, the concentration of the active ingredient becomes diluted below effective levels. Under these conditions, microorganisms have the ability to adapt or build up a tolerance to these antimicrobials. Highly resistant strains can develop that are immune to what was once an effective treatment dose.

Conventional (unbound) antimicrobials often can be very effective against specific types of microorganisms, but are generally limited in their ability to offer broad spectrum control. In other words, they may be effective against specific bacteria, but not all, or they may destroy all bacteria, but be ineffective against fungi, yeasts, or algae. The safety and toxicity of "unbound" antibacterial treatments vary considerably depending on the specific chemistry involved.

Bound Antimicrobials

Bound antimicrobial agents, like 3-trimethoxysilylpropyl dimethyloctadecyl ammonium chloride (SYLGARD Antimicrobial Treatment) manufactured by Dow Corning Corporation, remain chemically attached to the surface on which they are applied. They function by interrupting the organism's delicate cell membrane.[9] This prevents microorganisms from carrying on vital life processes. These antimicrobials kill organisms on contact and can do so again and again.

Since a "bound" antimicrobial is covalently and/or ionically bonded to surfaces, it does not diffuse or partition into the surrounding environment. An effective level of the material remains on the surface, and the adaptation process described earlier cannot and does not occur. The unique mechanism by which bound antimicrobials exhibit their activity permits them to effectively control a broad spectrum of microorganisms. Bacteria, molds, mildew, fungi, yeasts, and algae can all be controlled with this type of antimicrobial.[10]

BUILDING EVALUATIONS

Residential Study

Methodology

A total of 19 homes in the metropolitan area of Cincinnati were selected for the study, at least ten of which housed adolescent mold allergy sufferers. The homes were selected in conformance with the following criteria: (1) at least one family member had to be under the care of an allergist for at least 1 year and diagnosed as mold sensitive, (2) the attending allergist was asked to document clinical observations for at least 6 months, and (3) carpet and air conditioning were required in the main living areas of the home.

Prior to initiating the study, the following characteristics of each home were noted: (1) type, size, and age of home; (2) type of air conditioning; (3) presence and type of air filtration devices; (4) presence and type of other allergy control actions used in the home; and (5) characteristics of carpeting in the home as to age, amount, and wall-to-wall or area. The following parameters were recorded about the mold-sensitive occupants in each home: (1) age, (2) sex, (3) relative degree of severity in allergic responses, (4) other allergies, (5) current allergy therapy, and (6) name and length of time under the care of an allergist.

Testing

Two weeks prior to treatment standard plastic petri dishes (BBL) containing Sabauroud's Dextrose Agar were placed at floor level in random arrays (20 plates per home) throughout test zones. Plate locations, time, activity, and ambient conditions within zones were recorded.

Two weeks following treatment, petri dishes were placed at floor level in the pre-treatment locations. Post-treatment samplings were designed to replicate pre-treatment conditions as closely as possible. All plates were exposed for 1 h, sealed, and sent to the laboratory for incubation and enumeration using standard microbiological methods.

Participants were aware that they were part of a study but not informed regarding control or treated homes.

Results

Comparisons of total aeromicrobial gravity plate retrievals and percent changes before and after silane modified quaternary amines treatment can be seen in Figure 1.

Average total microbial retrievals in the homes prior to antimicrobial treatment of the carpet ranged from 6 colony forming units (CFUs) per plate to 42 CFUs per plate (Figure 1). After antimicrobial treatment, the average total microbial retrievals ranged from 1 CFU per plate to 20 CFUs per plate.

In 13 of the 19 homes (68%), greater than 50% reduction in total aeromicrobiological populations was shown following antimicrobial treatment of the carpeting.

Analysis of the symptomatic responses from the mold-sensitive occupants in the homes revealed that 19 of 24 (79%) recorded intermediate to significant improvement in their conditions. The improvements noted were fewer headaches, decreased congestion, better balance, decreased sinus problems, required medicine reduced or stopped, and an overall better feeling. The remaining five allergy sufferers recorded essentially no changes in their allergic symptoms. Three of the original study participants reported being ill with colds or other infections during the evaluation period, and the allergy-sufferer in the control house (#19) reported no change of condition. These four original participants are not included in the calculation above.

One year after the treatment was applied, participants were sent surveys to assess symptom scores, medication patterns, general health, and treatment value to allergics. Eighteen participants responded. These data are presented in Table 1.

Commercial Building Studies

Methodology

Studies on ten buildings from various geographical locations (Table 2) are reported in this paper. These buildings represent a wide array of structures and geographies. The common thread is the widespread reporting of SBS symptoms from the building occupants. Suspecting microbial involvement sourced from the environmental surfaces, microbial retrievals and mediation was undertaken.

This study was designed to determine gross variances of bioaerosol presence within large test areas. Gravitational sampling was utilized to provide broad aeromicrobiological profiles of test zones, thereby enabling a quantification of retrievals prior to and following treatment. Although the recovery of airborne agents, often in patterns that roughly parallel clinical events, has fostered widespread confidence in the validity of fallout techniques,[12] this retrieval method cannot be used to quantify changes in aerobiological densities. However, the repeated demonstration of statistically significant variances from a sufficiently high number of sampling locations provides confidence in identifying an event as causal and allows for gross comparisons at specific sample sites.

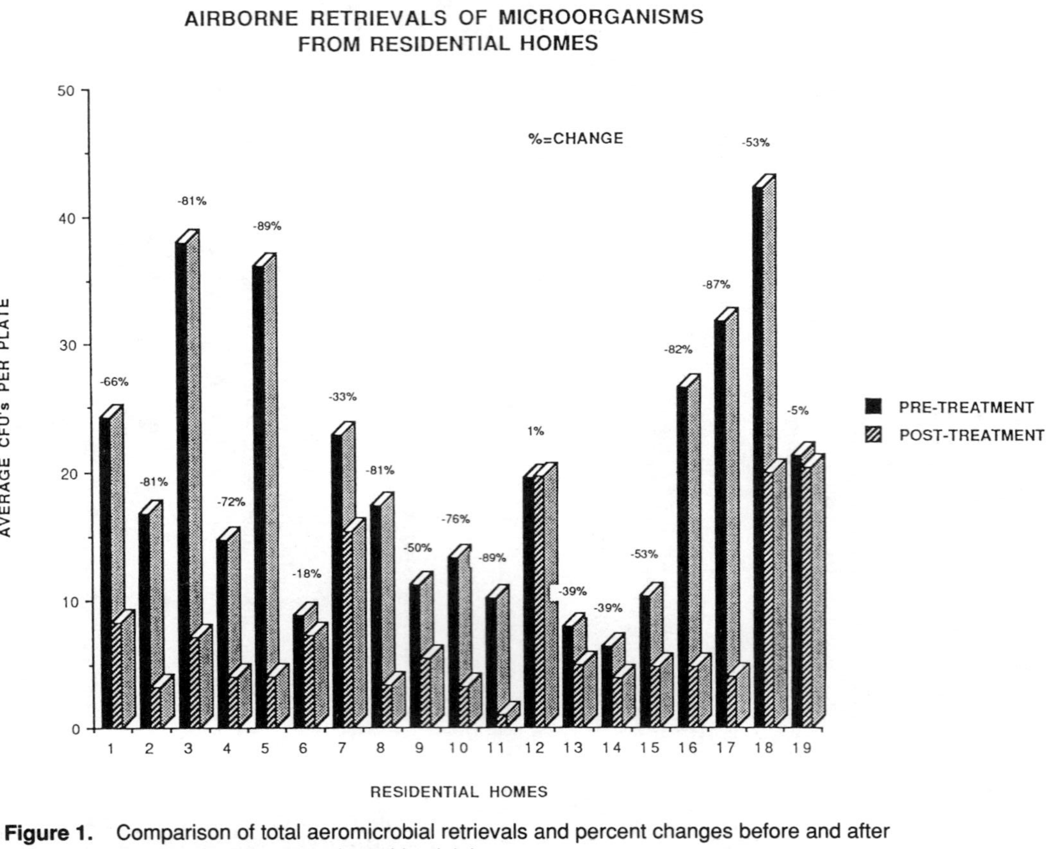

Figure 1. Comparison of total aeromicrobial retrievals and percent changes before and after antimicrobial treatment in residential homes.

Table 1. 1988 Residential Mold Allergy Study Survey Results 1 Year After Treatment

Symptom Scores During Study		
	Pre-Treatment	**1 Year After**
Mild	0	8
Moderate	10	7
Mod-Severe	3	2
Severe	5	1
Reported Changes in General Health 1 Year After Treatment		
Improved	10	
Worse	1	
Unchanged	7	
Alteration of Medications During Study		
Increased dosage or frequency	0	
Decreased dosage	1	
Decreased frequency	1	
Decreased dose and frequency	1	
Decreased dose, frequency and type	4	
Unchanged	11	
Reported Usefulness of Treatment to Allergics		
Should be available to allergics	18	
Not beneficial to allergics	0	

Table 2. Commercial Building Studies — Building Codes

Number	Type	Location
1	School	Alexandria, KY
2	Office building/print shop	St. Petersburg, FL
3	Office building	Rochester, NY
4	Condominiums	Keystone, CO
5	Office building	Clearwater, FL
6	Office building	Clearwater, FL
7	Office complex	Clearwater, FL
8	Office building	Miami, FL
9	Office building	Tampa, FL
10	Office building	Cincinnati, OH

Treatment

An aqueous solution of 3-trimethoxysilylpropyl dimethyloctadecyl ammonium chloride was applied to dry carpeting in accordance with the manufacturer's specifications.[11] Carpeting was not cleaned prior to antimicrobial applications. Building occupants in six of the buildings were not aware of any remediation activities. Although samplings were performed during normal work hours, application of the treatment was performed at night or on weekends without their knowledge.

Testing

Two weeks prior to treatment, standard plastic petri dishes (BBL) containing Sabauroud's Dextrose Agar were placed at floor level in random arrays (14 to 50 sites per building) throughout test zones. Plate locations, time, activity, and ambient conditions within zones were recorded.

Two weeks following treatment, petri dishes were placed at floor level in the pre-treatment locations. Post-treatment samplings were designed to replicate pre-treatment conditions as closely as possible. All plates were exposed for 1 h, sealed and sent to the laboratory for incubation and enumeration using standard microbiological methods.

Results

Data and observations of ten buildings are reported in this paper. These are representative of all buildings we have investigated, both in quantification of variation and clinical observations of occupant response. The percent variation of each building following treatment of carpeting is shown in Figure 2. These averages are derived by dividing the total number of colonies retrieved by the number of plate sites.

The variation between pre-treatment and post-treatment retrieval averages range between 71 and 98%. Within this group of buildings, two showed greater than 90% change, nine showed greater than 80% change, and all showed greater than 70% change.

The data in Figure 3 are representative of patterns observed in the ten buildings in this study. Note the pre-treatment variances representing a range of from 2 CFUs per plate to 156 CFUs per plate whereas the post-treatment retrieval counts range only from 0 CFUs per plate to 4 CFUs per plate. This stabilization of the aeromicrobiological retrievals is noteworthy along with the consistently effective reduction in numbers retrieved.

The clinical profiles of building occupants within the commercial buildings were evaluated during the 12 months following treatment. No changes were reported or observed in any of the buildings. During the second year following treatment, aerobiological samplings were performed at five of the buildings in conformance with the initial and post-treatment sampling criteria. The retrieval averages are presented in Table 3 and reveal aeromicrobiological profiles in ranges consistent with post-treatment averages.

In the ten investigations in this report of BRI/SBS within a large diversity of building designs and geographies, symptomatic improvement was uniformly reported from workers and reduction of microbioaerosol levels were observed after treatment of the carpeting with the silane modified quaternary amine. While these data are not conclusive, it challenges us to dislodge traditional perceptions and expand our research efforts to better understand the short- and long-term health effects that result from exposure to microbiological pollutants in the workplace.

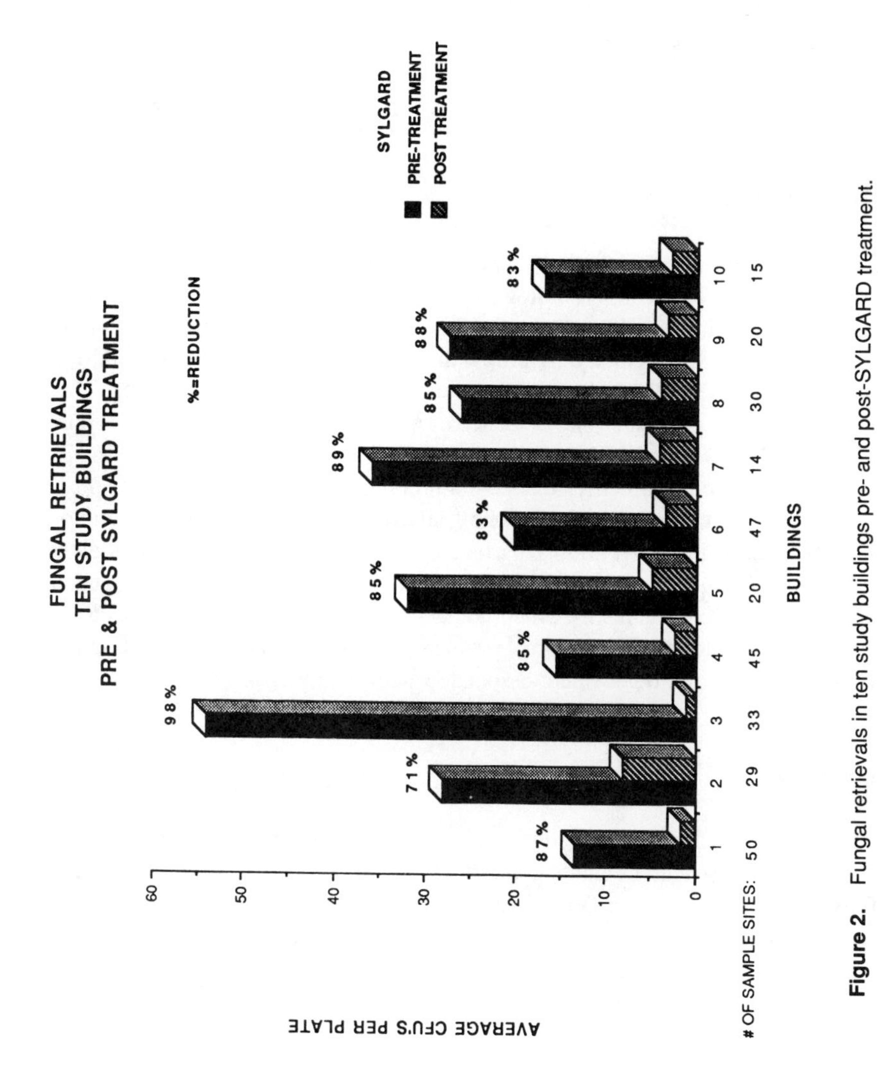

Figure 2. Fungal retrievals in ten study buildings pre- and post-SYLGARD treatment.

Table 3. Follow-Up Sampling in Five Buildings During 2nd Year After Treatment

Building	Average CFUs Retrieved per Plate		
	Pre-Treatment	Post-Treatment	2nd Year
1	13.4	1.7	3.6
3	54.0	1.0	1.1
6	20.3	3.5	4.1
9	27.4	3.3	3.5
10	17.0	2.9	2.8

CONCLUSIONS

These studies provide data that support previous claims that carpeting contributes substantively to aeromicrobiological presence within buildings. It is the first attempt to determine whether or not microbioaerosol presence can be regulated by the application of 3-trimethoxysilylpropyl dimethyloctadecyl ammonium chloride to carpeting and be reflected in agar plate retrievals and in human response. Thus, our investigations present strong evidence of microbial involvement in the acquisition of BRI/SBS and reveals an effective remediation tool in the form of the 3-trimethoxysilylpropyl dimethyloctadecyl ammonium chloride.

The durable attachment of 3-trimethoxysilylpropyl dimethyloctadecyl ammonium chloride to interior building surfaces clearly reduces aeromicrobiological densities. The unique functionality of these activated surfaces enable the extended destruction of microorganisms which contact them. This technology provides a useful tool for dealing with microbial problems on surfaces and for mediating the morbidity, odors, and defacement associated with microorganisms.

ACKNOWLEDGMENTS

The authors wish to acknowledge and thank Rich Gettings for his microbiological support in these studies and to thank Julie Clapper for her help in following the subject responses in the residential study.

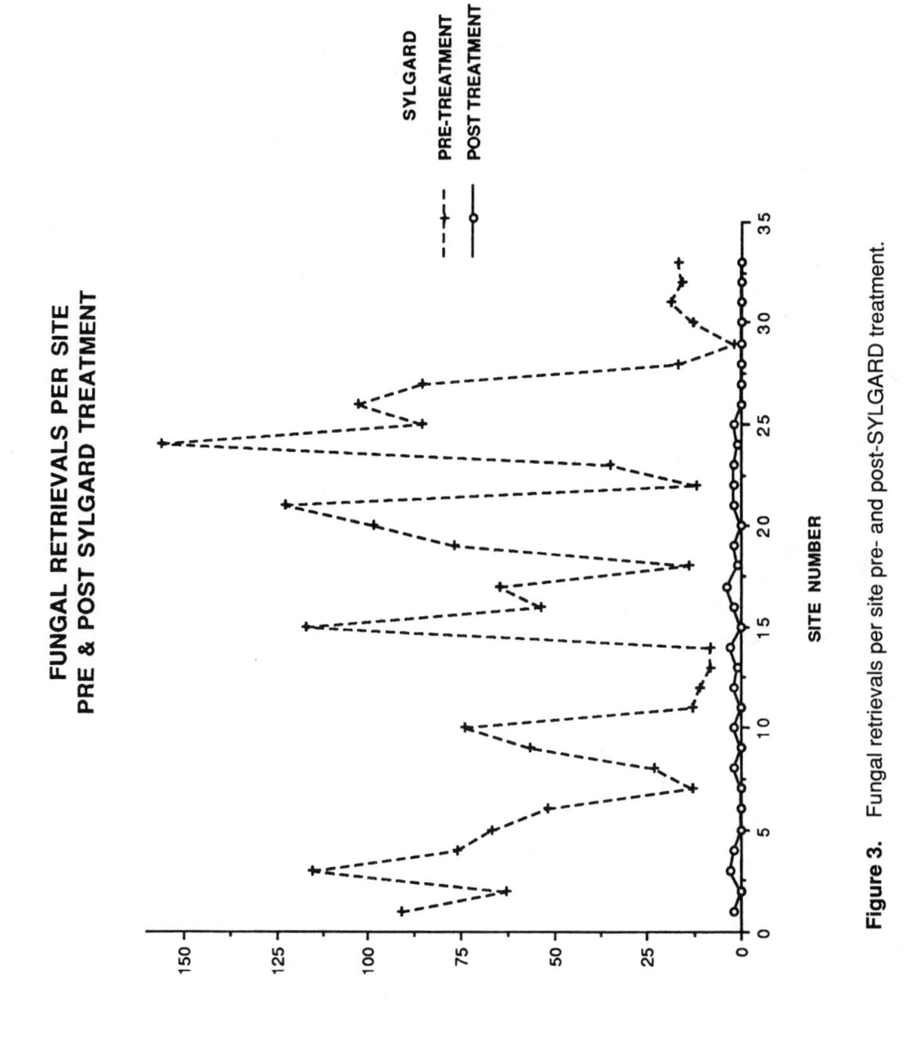

Figure 3. Fungal retrievals per site pre- and post-SYLGARD treatment.

REFERENCES

1. WHO Reports, *Indoor Air Pollutants: Exposure and Health Effects,* WHO Meeting in Nordlingen 1982. Eur. Report No. 78, (1983).
2. American Society of Heating, Refrigeration, and Air Conditioning Engineers, *Indoor Air Quality Position Paper,* (Atlanta: ASHRAE, August 11, 1987).
3. National Institute for Occupational Safety and Health (NIOSH), *Guidance for Indoor Air Quality Investigations,* (January 1987).
4. WHO Reports, *Biological Contaminants in Indoor Air.* WHO Meeting in Rautavaara 1988. Eur. Report
5. Sorenson, W.G., "Mycotoxins as Potential Occupational Hazards," Annual Meeting of the Society for Industrial Microbiology, (Seattle, WA: August 1989).
6. Burge, H.A., Annual meeting of the Academy of Allergy and Immunology, (Anaheim, CA: March 16, 1988).
7. Brundage, J.F., "Building Associated Risk of Febrile Acute Respiratory Diseases and Army Trainees," *JAMA,* 259:14 (1988).
8. White, W.C. and Gettings, R.L., "Evaluating the Antimicrobial Properties of Silane Modified Surfaces." in *Silanes, Surfaces, and Interfaces,* Leyden, D.E., (Ed.), (New York: Gordon and Breach Science Publishers, 1986).
9. White, W.C. and Gettings, R.L., "Evaluating the Antimicrobial Properties of Silane Modified Surfaces." in *Silanes, Surfaces, and Interfaces,* Leyden, D.E., (Ed.), (New York: Gordon and Breach Science Publishers, 1986).
10. Speier, J.L. and Malek, J.R., "Destruction of Microorganisms by Contact with Solid Surfaces," *J. Colloid Interface Sci.,* 89(1):68–76 (1982).
11. Dow Corning Corporation, *SYLGARD Treatment Manual,* Form No. 24-264a-88, (1988).
12. Soloman, W.R., "Sampling Techniques for Airborne Fungi," in *Mould Allergy.* (Philadelphia: Lea and Febiger, 1984) p. 46.

CHAPTER 6

Indoor and Outdoor Concentrations of Organic Compounds Associated with Airborne Particles: Results Using a Novel Solvent System

Datta V. Naik, Charles J. Weschler, and Helen C. Shields

ABSTRACT

An extraction method, using a mixed solvent, has been developed for the identification of semivolatile organic compounds (SVOCs) associated with airborne particles. The procedure yields a single extract, suitable for gas chromatographic analysis, that contains both polar and nonpolar compounds. The solvent is a one-to-one mixture of cyclohexane and 1-chlorobutane. The liquids are miscible and have virtually identical boiling points, which permits refluxing as well as evaporative concentration. The extraction procedure, followed by gas chromatograph/mass spectrometer (GC/MS) analysis, has been used to examine SVOCs associated with size-fractionated airborne particles. The particles were collected indoors and outdoors, for a 6-week period, at a site in Neenah, Wisconsin. The procedure adequately extracts the major organic compounds associated with these particles. The results indicate that there are strong indoor sources for certain branched and *n*-alkanes, fatty acids, and phthalate esters at the Neenah site. It is also apparent that the organic compounds associated with indoor coarse particles have indoor sources. This study reaffirms the earlier observation that partitioning between the vapor phase and the surface of airborne particles is more pronounced indoors than outdoors.

INTRODUCTION

Semivolatile organic compounds (SVOCs), especially those associated with fine airborne particles, can penetrate deep into the lungs and remain in contact with lung tissue for an extended period. Consequently, such compounds are a potential threat to human health. SVOCs can also contribute to failures in sensitive electronic equipment. Examples of such failures include promoting arcing between contacts of switches and relays leading to increased contact erosion, forming resistive films on precious metal contacts, and contributing to computer head crashes and disc drive failures.

Numerous studies have examined SVOCs associated with outdoor airborne particles (see, for example, References 1 to 8 and references therein). Still other studies have analyzed SVOC associated with indoor airborne particles and contrasted the results with those from outdoor samples.[9-12] In the majority of these studies the organic compounds were extracted from the collected particles using a variety of solvents. Chromatographic procedures were then applied to the resulting extracts. The collection of particles and the chromatographic analysis are usually straightforward. The extraction step, however, can be time consuming and subject to contamination and losses. For example, Daisy and co-workers[5,6] have studied SVOC associated with particles using a sequential three-solvent extraction (solvents of increasing polarity: cyclohexane, dichloromethane, and acetone). Weschler et al. have employed methanol,[9] toluene,[10] petroleum ether,[12] and a thermal desorption method[10,12] to examine SVOC. Multiple solvent extraction is a lengthy process, and each extraction step can lead to loss of analytes, potential contamination, and artifacts. Thermal desorption is relatively fast but gives lower signal-to-noise ratio and consequently poor detection limit. In addition, the use of a low desorption temperature (250 to 300°C) makes it difficult to quantify high molecular weight compounds. Higher desorption temperatures may be used, but that can lead to decomposition of unstable compounds that may be present.

Solvent extraction is still a preferred method for the analysis of SVOC associated with particles as it allows for higher sensitivity. An extraction procedure that yields a single extract is desirable, as this avoids problems associated with multiple solvent extractions. In addition, use of a single extract facilitates subsequent chromatographic analysis.

In this study we report a simple extraction procedure using a mixed solvent system that yields a single extract, containing both polar and nonpolar compounds, suitable for chromatographic analysis. This method, followed by gas chromatograph/mass spectrometer (GC/MS) analysis, has been used to examine SVOCs associated with size fractionated airborne particles collected indoors and outdoors at a site in Neenah, Wisconsin.

EXPERIMENTAL

Sampling

Size fractionated aerosol samples were collected simultaneously indoors and outdoors at a telephone switching office in Neenah during the 6-week period, January 25 to March 8, 1989. Automatic dichotomous samplers (Anderson Sierra Model No. 245) were used to collect coarse particles (2.5 to 15 μm) and fine particles (< 2.5 μm) on Teflon membrane filters (37-mm diameter). Prior to their use, the filters were subjected to 6 successive ultrasonic washes (30 min each), first with methanol, then water, then methanol, etc., to remove any contaminants that may be present on virgin filters. The dichotomous sampler was programmed to use a pair of filters (1 coarse, 1 fine) per week. The details of the sampling systcm are given elsewhere.[13]

Procedure

A solvent system was sought in which a nonpolar solvent would be used to extract nonpolar organic compounds from particles; a polar solvent would then be added to the nonpolar extract to remove remaining polar compounds from the particles. The second solvent has to be miscible with the first, and on mixing should raise the polarity of the mixture enough to extract the polar compounds. The increase in solvent polarity should not be so much as to precipitate the nonpolar compounds already extracted. Additionally, since the extract will then be subjected to evaporative condensation from a few milliliters to a few microliters, this evaporation should not significantly enrich either of the solvent components in the extract.

Cyclohexane and 1-chlorobutane were chosen as the two solvents that meet the criteria mentioned above. Cyclohexane (bp 81°C) is frequently used as a solvent for extraction of nonpolar compounds (e.g., Daisey et al.[5,6]) The boiling point of 1-Chlorobutane (78°C) is close to that of cyclohexane, and the two solvents are miscible. The dielectric constant of 1-chlorobutane (7.4) is similar to that of dichloromethane (8.9) — a solvent commonly used for extraction of polar organic compounds.

The extraction procedure involved sonicating the Teflon membrane filters (six coarse filters or six fine filters), containing particles collected over the 6-week period, in 1.00 ml of cyclohexane for 2 h. This was followed by 12 to 15 h of soaking, and then 1.00 ml of 1-chlorobutane was added. The mixture was again sonicated for 2 h followed by an additional 24 h of soaking. The Teflon filters were then removed and the extract condensed to 10 to 50 μl in a fume hood at room temperature. Aliquots of 1 μl were injected into the GC/MS. Six pre-washed blank filters were also carried through the same extraction procedure and analysis.

Organic compounds contained in the extract were analyzed using a Hewlett Packard 5995C (GC/MS). The separation was achieved with a 12 m × 0.2 mm, cross-linked dimethyl silicone, fused silica capillary column, and a helium flow rate of 1 ml/min in split operation. The GC injection port was kept at 240°C, and the oven was programmed to start at 50°C for 2.6 min, followed by a temperature rise to 300°C at a rate of 8°C/min.

The response of the mass spectrometer was calibrated using an internal standard (d_{10}phenanthrene) and pure analytes (*n*-alkanes, phthalate esters, fatty acids) with selected ion monitoring (SIM) techniques. The ions monitored during a GC/MS run are listed in Table 1. The specific ions used for quantitation of respective

Table 1. Ions Monitored During GC/MS Analysis Using the SIM Technique

m/e Monitored*	Compound(s)
57	*n*-Alkanes with C > 34
73	Fatty acids
85	*n*-Alkanes with C > 34
127	*n*-$C_{17,23,33,34}$ and br-$C_{30\text{-}32}$
141	Branched alkanes
<u>149</u>	Phthalate esters
160,**188**	d_{10}-Phenanthrene (internal standard)
223	Dibutyl phthalate
256	Palmitic acid
279	Di(2-ethylhexyl)phthalate
284	Stearic acid
293	Dinyl phthalate
307	Didecyl phthalate
338	*n*-C_{24}
352	*n*-C_{25}
366	*n*-C_{26}
380	*n*-C_{27}
394	*n*-C_{28}
408	*n*-C_{29}
422	*n*-C_{30}
436	*n*-C_{31}
450	*n*-C_{32}
464	*n*-C_{33}
478	*n*-C_{34}

* The bold m/e peaks were used for the quantitation of the respective compounds.

analytes appear in the table as underscored figures. Samples were quantified by comparison of mass spectral response of selected ion for each analyte to that of the same ion for the respective standard, both normalized to the molecular ion (188) of the internal standard. A few analytes (br-alkanes, *n*-alkanes with C>34, dinonyl and didecyl phthalate) for which standards were not used were estimated by assuming their MS response to be similar to other structurally related standards. Recoveries of standards were determined by spiking a composite of six pre-washed filters with the standards and subjecting these filters to the extraction and analysis. The recoveries were quantitative for *n*-C_{17}, palmitic acid, and stearic acid. For other standards the recoveries were between 59 and 99%; the appropriate recovery factors were used in the quantitation step. No blank corrections were needed since the GC/MS analysis (SIM mode) of the extract of the six pre-washed blank filters did not detect any of the analytes of interest. All chemicals (solvents and standards) were of HPLC of Gold Label grade, and were obtained from the Aldrich Chemical Co. and Supelco Inc. The solvents were subjected to GC/MS analysis after evaporative condensation of 2.0 ml of the 1:1 mixture (cyclohexane/1-chlorobutane) to a few microliters. No contaminants were detected.

RESULTS AND DISCUSSION

A 1:1 mixture of the two solvents was found to reflux at 77°C. The solvent composition for 2.00 ml of the 1:1 mixture was studied as a function of evaporation, at room temperature, using a GC. The composition changed gradually with evaporation, progressing towards slight enrichment of cyclohexane. When 2.00 ml of the mixture was evaporated down to 10 ml, the final composition was 54.2% cyclohexane and 45.8% 1-chlorobutane — not a large enough enrichment to cause precipitation of extracted compounds.

A trial analysis was conducted on an aggregate of six 1-week outdoor coarse samples (collected at Neenah) using the above extraction procedure and GC parameters. The mass spectrometer was operated to present a total ion current profile ranging in mass from 40 to 450. The total ion chromatogram of the trial extract is presented in Figure 1. The analytical procedure was found to extract both polar and nonpolar compounds of interest. For this particular sample (coarse airborne particles) n-alkanes as large as C_{43} were detected.

Given the very small amounts of SVOC present (0.02 to 13 ng/m^3), all the GC/MS analyses on real samples were performed using SIM techniques. The SIM method increases the signal-to-noise ratio and improves the detection limit by as much as 1000-fold compared to total ion current techniques.[14]

The average outdoor and indoor concentrations of the major SVOC associated with fine and coarse airborne particles, collected over the 6-week period (January 25 to March 8, 1989), are summarized in Table 2 and Table 3. The average mass concentration of combined fine and coarse airborne particles outdoors was 28,200 ng/m^3, and that indoors was 1290 ng/m^3. During the sampling period the building fans were operated continuously and the air was circulated through filters with an ASHRAE Dust Spot Rating of 85%. The significantly lower concentration of total indoor airborne particles (<5% of the outdoor level), and extremely low concentration of indoor coarse particles (<0.7% of the outdoor level) are attributable to the efficient air filtration at this site coupled with the relatively small number of indoor sources generating airborne particles.

The major nonpolar SVOC detected in the outdoor and indoor samples (Table 2) are the n-alkanes, C_{26} to C_{33}, and the branched alkanes, C_{30} to C_{32}. The combined (fine plus coarse) concentrations for the nonpolar compounds range between 0.3 and 2.4 ng/m^3 in the outdoor samples, and 0.4 and 13 ng/m^3 in the indoor samples. The most abundant (>2.2 ng/m^3) alkanes associated with the outdoor particles are n-C_{29} and n-C_{31}. Among the nonpolar SVOC detected indoors, the abundant (>2.2 ng/m^3) compounds are n-C_{29}, n-C_{31}, n-C_{33}, br-C_{31}, and br-C_{32} with the concentration of n-C_{31} the highest (13 ng/m^3). More will be said about n-C_{31} below. As a class the n-alkanes display an odd/even carbon number predominance (Figure 2) in indoor as well as outdoor samples. This suggests that a significant fraction of these compounds are derived from vascular plant waxes.[4,8] Common indoor sources for the n-alkanes include sealants, lubricants, and polishes.[15] In addition to the compounds listed in Table 2, the chromatograms also displayed a "hump" starting at

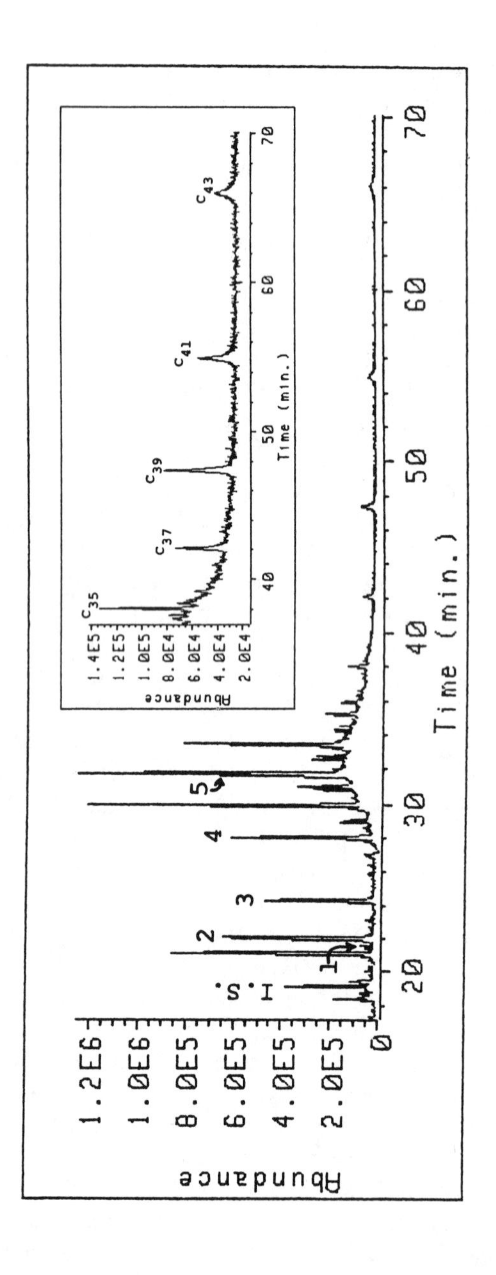

LEGEND: I.S.=Int. std., 1=DBP, 2=Palm. Acid,
3=Stear. acid, 4=DEHP, 5=DDP.
Unlabeled=Nonpolar Compounds.

Figure 1. Total ion chromatogram of a trial extract of an outdoor coarse sample.

Table 2. Average Outdoor and Indoor Concentrations for Major Nonpolar Organic Compounds Associated with Airborne Particles at Neenah, Wisconsin

Compound	Outdoor ng/m^3		Indoor, ng/m^3	
	Fine	Coarse	Fine	Coarse
Airborne particles	14300	13900	1200	90
n-C_{26}	0.21	0.10	0.42	0.01
n-C_{27}	0.39	0.51	1.20	n.d.
n-C_{28}	0.42	0.10	0.40	0.02
n-C_{29}	1.29	0.96	2.53	n.d.
n-C_{30}	0.76	0.96	1.24	n.d.
br-C_{30}	0.9	n.d.	1.9	n.d.
n-C_{31}	1.78	0.62	12.94	n.d.
br-C_{31}	0.7	n.d.	4.0	n.d.
n-C_{32}	0.46	0.14	1.50	n.d.
br-C_{32}	0.6	n.d.	4.9	n.d.
n-C_{33}	1.62	n.d.	4.06	n.d.

Table 3. Average Outdoor and Indoor Concentrations for Major Polar Organic Compounds Associated with Airborne Particles at Neenah, Wisconsin

Compound	Outdoor, ng/m^3		Indoor, ng/m^3	
	Fine	Coarse	Fine	Coarse
Airborne particles	14300	13900	1200	90
Dibutylphthalate	0.08	0.05	0.04	0.02
Di(2-ethylhexyl)phthalate	1.4	0.35	2.5	0.5
Palmitic acid	10.6	2.8	1.2	1.1
Stearic acid	9.8	1.1	1.9	1.0

about n-C_{26} and extending to n-C_{36}. This broad band is due to the elution of high molecular weight branched and substituted cyclic alkanes with boiling points so close that they could not be separated with the capillary column used in this study. Such a hump is commonly referred to as an unresolved complex mixture (UCM). Other studies[7,16] have reported similar high-boiling UCMs in extracts of particulate phase organic compounds and have ascribed the source to be anthropogenic emissions (primarily fossil fuel combustion). Although the UCM is apparent in the total ion chromatograms (see Figure 1), it does not interfere with SIM analyses due to the selective nature of the latter technique.

n-C_{31} has been detected in high concentration in other indoor environments (e.g., Wichita, KS and Lubbock, TX.[10,12] Its unusually high indoor concentration at each of these sites suggests a strong, yet fairly common, indoor source. A possible candidate is smoking. In chamber studies, Eatough et al. have identified n-C_{31} as one of the major particle-phase compounds found in environmental tobacco smoke (ETS).[17] It is the only linear alkane so identified. Although smok-

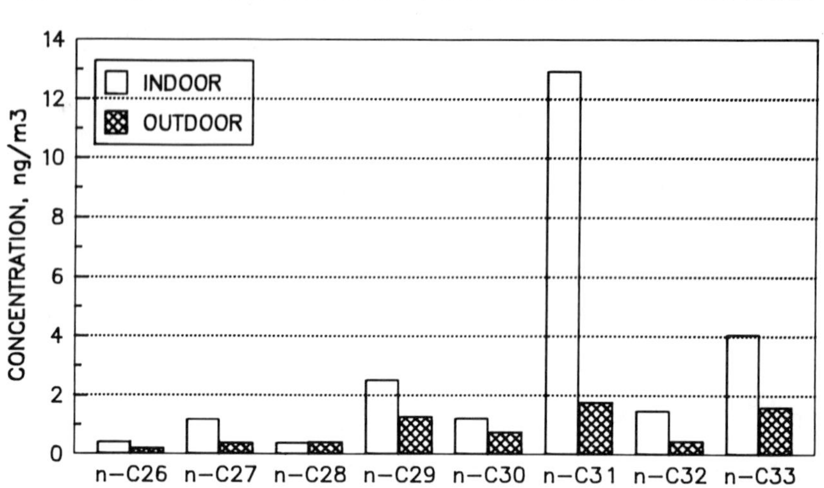

Figure 2. Distribution of selected *n*-alkanes associated with indoor and outdoor fine particles.

ing is prohibited in telephone switching areas, the Neenah office does have a "breakroom" where employees are permitted to smoke. A certain fraction of the air from this room is recirculated throughout the office by the building's air handling system. Fine-mode potassium (K) has also been identified as a potential tracer of ETS.[17] Indeed, in other studies of telephone switching offices (including Wichita and Lubbock), elevated levels of fine-mode potassium have been ascribed to smoking.[18,19] In a separate study, the water-soluble salts associated with fine and coarse particles from Neenah have been measured.[20] The concentration of indoor fine-mode potassium averaged 9 ng/m^3. Hence the ratio, by mass, of the average indoor concentration of n-C_{31} to K at Neenah is (12.9/9) = 1.4. The ratio, by mass, of n-C_{31} to K in Eatough's chamber studies was 1.2.[17] This simple analysis does not take into account the relative contribution of outdoor sources to the indoor concentrations of these species. However, it does show that smoking is a reasonable explanation for elevated indoor levels of both n-C_{31} and K.

The major polar compounds detected in the outdoor and indoor samples (Table 3) are dibutyl phthalate, di(2-ethylhexyl)phthalate (DEHP), and fatty acids (palmitic and stearic acids). Phthalate esters are frequently used as plasticizers,[15] and their association with airborne particles is expected from the widespread use of plastics. Palmitic and stearic acids have many natural sources, including vascular plant waxes,[4,21] and are also present in a variety of commercial products.[15]

The indoor/outdoor ratios (I/O) for selected nonpolar and polar compounds associated with the airborne particles collected over the 6-week period are listed in Table 4. The average I/O values for the mass of the aerosol particles (0.08 for

Table 4. Indoor/Outdoor Ratios for Selected Nonpolar and Polar Compounds

Compound	I/0 fine	I/0 coarse
Airborne Particles	0.08	0.006
n-C_{27}	3.1	—
n-C_{28}	0.9	0.2
n-C_{29}	2.0	—
n-C_{30}	1.6	—
br-C_{30}	2.1	—
n-C_{31}	7.3	—
br-C_{31}	5.7	—
n-C_{32}	3.3	—
br-C_{32}	8.2	—
n-C_{33}	2.5	—
Dibutylphthalate	0.5	0.4
Di(2-ethylhexyl)phthalate	1.7	1.4
Palmitic acid	0.1	0.4
Stearic acid	0.2	0.9

fine mode and 0.006 for coarse mode) provide a framework in which the I/O relationships for the SVOC can be considered. If the I/O value for a fine-mode SVOC is significantly greater than 0.08 (0.006 in the case of a coarse-mode SVOC), then that compound is assumed to have a strong indoor source.

For most of the coarse-mode alkanes, I/O values could not be calculated since their indoor concentrations were below our detection limit (0.02 ng/m^3). Apparently, indoor sources of these alkanes in the coarse mode are insignificant. For all of the fine-mode alkanes, the I/O values indicate significant indoor sources. The data suggest stronger indoor sources for n-C_{27}, n-C_{31}, n-C_{32}, n-C_{33}, br-C_{31}, and br-C_{32} than for n-C_{28}, n-C_{29}, n-C_{30}, and br-C_{30}. As mentioned above, common indoor sources for the alkanes include waxes, polishes, and lubricants.[15]

Comparisons of the I/O values for the phthalate esters with the corresponding values for the airborne particles indicate strong indoor sources of dibutyl phthalate and DEHP in the fine as well as in the coarse mode. Dibutyl phthalate is frequently used as a plasticizer in floor polishes. DEHP is used to plasticize vinyl products and can be found in everything from floor tiles to vinyl binders. The abrasion of materials such as floor tiles could contribute to the relatively high concentration of these SVOC in the coarse mode. Comparisons of the I/O values for the fatty acids with the corresponding values for the airborne particles suggest that there are also indoor activities which directly introduce course-mode particles containing these compounds. Indoor fatty acid sources are comparatively less important in the fine mode. Potential indoor sources of the fatty acids include detergents, lubricants, waxes, and polishes.[15] Further analyses that may clarify the source(s) of these coarse-mode fatty acids are in progress.

The fine/coarse ratios (F/C) for selected nonpolar and polar compounds are

Table 5. **Fine/Coarse Ratios for Selected Nonpolar and Polar Compounds**

Compound	F/C Outdoor	F/C Indoor
Airborne Particles	1.03	13.8
n-C_{27}	0.8	>100
n-C_{28}	4.2	20
n-C_{29}	1.3	>200
n-C_{30}	0.8	>100
br-C_{30}	>45	>90
n-C_{31}	2.9	>1000
br-C_{31}	>35	>200
n-C_{32}	3.3	>100
br-C_{32}	≥30	≥245
n-C_{33}	>150	>300
Dibutylphthalate	1.6	2.0
Di(2-ethylhexyl)phthalate	4.0	5.2
Palmitic acid	3.8	1.1
Stearic acid	8.8	1.8

listed in Table 5. Comparing the amounts of a given SVOC in the two modes with the corresponding mass ratio of fine and course particles can reveal the manner in which the organic compounds are associated with the particles. The surface area per unit mass of fine particles is normally larger than that of coarse particles by an order of magnitude or more. A high F/C value for a SVOC, scaling with the ratio of the fine/coarse particle surface area, and not with the ratio of fine/coarse particle mass, suggests that the organic compound is primarily adsorbed on the surface of the particles. Conversely, a F/C value for a SVOC comparable to the corresponding F/C value for particle mass indicates that the compound is integrally associated with the airborne particles (i.e., the compound occurs throughout the particles rather than just on the particle surface).

The outdoor F/C values of n-C_{27}, n-C_{28}, n-C_{29}, n-C_{30}, n-C_{31}, and n-C_{32} range from 0.8 to 4.2 compared to the corresponding outdoor particle mass ratio (F/C) of 1.03. The outdoor F/C values for br-C_{30}, br-C_{31}, br-C_{32}, and n-C_{33} are all larger than 30, indicating that, compared to the other alkanes (n-C_{27} to n-C_{32}), a significantly greater percentage of these compounds are adsorbed on the surface of outdoor particles.

The indoor F/C values of all the hydrocarbons (except n-C_{28}) listed in Table 5 are estimated to be greater than 90. The comparison of these values with the indoor particle mass ratio (F/C) of 13.8 suggests that these hydrocarbons are primarily adsorbed on the surface of the indoor airborne particles. These observations imply that for all the alkanes (except n-C_{28}), partitioning between the vapor phase and the surface of airborne particles occurs to a greater extent indoors than outdoors. This is reasonable since it is easier to achieve an elevated vapor phase concentration of SVOC in the limited volume of an indoor environment than in the outdoor environment. Elevated concentrations are even more likely when ventilation rates are reduced through energy conservation measures.

The F/C values for the polar compounds range from 1.6 to 8.8 for outdoor samples and from 1.1 to 5.2 for indoor samples. Comparison of these values with the corresponding particulate ratio (F/C) suggest that partitioning of these compounds is relatively unimportant in both outdoor and indoor environments, and that they tend to be integrally associated with the airborne particles.

CONCLUSIONS

A mixed solvent (1:1 cyclohexane/1-chlorobutane) extraction procedure has been developed which adequately extracts polar and nonpolar SVOC associated with airborne particles. The results of the analysis of 6-week samples of coarse and fine particles collected indoor and outdoor at the Neenah site indicate that strong indoor sources exist for n-C_{27}, n-C_{29}, n-C_{31}, n-C_{32}, n-C_{33}, br-C_{30}, br-C_{31}, and br-C_{32}, and for dibutyl phthalate, DEHP, palmitic acid, and stearic acid. This study reaffirms the earlier observation[12] that partitioning between the vapor phase and the surface of airborne particles is more pronounced indoors than outdoors.

ACKNOWLEDGMENTS

We thank F. D. Hileman (Monsanto) for suggesting 1-chlorobutane as a solvent and A. W. Billstrand (Wisconsin Bell) for assistance throughout the sampling program at Neenah. DVN gratefully acknowledges support from the Bellcore-Monmouth College Cooperative Summer Faculty Research Program, and Monmouth College Grant-in-Aid for Creativity.

REFERENCES

1. Ketseridis, G., Hahn, J., Jaenicke, R., and Junge, C., "The organic constituents of atmospheric particulate matter," *Atmos. Environ.,* 10:603–610 (1976).
2. Eichmann, R., Neuling, P., Ketseridis, G., Hahn, J., Jaenicke, R., and Junge, C., "*n*-Alkane Studies in the Troposphere — I. Gas and Particulate Concentrations in North Atlantic Air," *Atmos. Environ.,* 13:587–599 (1979).
3. Giam, C.S., Atlas, E., Chan, H.S., and Neff, G.S., "Phthalate Esters, PCB and DDT Residues in the Gulf of Mexico Atmosphere," *Atmos. Environ.* 14:65–69 (1980).
4. Gagosian, R.B., Peltzer, E.T., and Zafiriou, O.C., "Atmospheric Transport of Continentally Derived Lipids to the Tropical North Pacific," *Nature (London),* 291:312–314 (1981).
5. Daisy, J.M., Hershman, R.J., and Kneip, T.J., "Ambient Levels of Particulate Organic Matter in New York City in Winter and Summer," *Atmos. Environ.,* 16:2161–2168 (1982).

6. Daisey, J.M., Morandi, M., Lioy, P.J., and Wolff, G.T., "Regional and Local Influences on the Nature of Airborne Particulate Organic Matter at Four Sites in New Jersey during Summer 1981," *Atmos. Environ.,* 18:1411–1419 (1984).
7. Doskey, P.V. and Andren, A.W., "Particulate- and Vapor-Phase *n*-Alkanes in the Northern Wisconsin Atmosphere," *Atmos. Environ.,* 20:1735–1744 (1986).
8. Simoneit, B.R.T. and Mazurek, M.A., "Organic Tracers in Ambient Aerosols and Rain," *Aerosol Sci. Technol.,* 10:267–291 (1989).
9. Weschler, C.J., "Characterization of Selected Organics in Size-Fractionated Indoor Aerosols," *Environ. Sci. Technol.,* 14:428–431 (1980).
10. Weschler, C.J., "Indoor-Outdoor Relationships for Nonpolar Organic Constituents of Aerosol Particles," *Environ. Sci. Technol.,* 18:648–652 (1984).
11. Lioy, P.J., Avdenko, M., Harkov, R., Atherholt, T., and Daisey, J.M., "An Indoor/Outdoor Study of Inorganic and Organic Particulate Matter and Particulate Mutagenicity, *JAPCA,* 35:653–657 (1985).
12. Weschler, C.J. and Fong, K.L., "Characterization of Organic Species Associated with Indoor Aerosol Particles," *Environ. Inter.,* 12:93–97 (1986).
13. Weschler, C.J., Shields, H.C., Kelty, S.P., Psota-Kelty, L.A., and Sinclair, J.D., "Comparison of Effects of Ventilation, Filtration, and Outdoor Air on Indoor Air at Telephone Office Buildings: A Case Study," in ASTM STP, 1002: *Design and Protocol for Monitoring Indoor Air Quality,* Nagda, N.L. and Harper, J.P. (Eds.), (Philadelphia: American Society for Testing and Materials, 1989). pp 9–34.
14. Budde, W.L. and Eichelberger, J.W., *Organics Analysis Using Gas Chromatography/Mass Spectrometry.* (Ann Arbor: Ann Arbor Science, 1979). p. 242.
15. Gosselin, R.E., Smith, R.P., and Hodge, H.C., *Clinical Toxicology of Commercial Products,* 5th ed., (Baltimore: Williams & Wilkins, 1984).
16. Simoneit, B.R.T., Mazurek, M.A., and Cahill, T.A., "Contamination of the Lake Tahoe Air Basin by High Molecular Weight Petroleum Residues," *JAPCA,* 30:387–390 (1980).
17. Eatough, D.J., Benner, C.L., Tang, H., Landon, V., Richards, G, Caka, F.M., Crawford, J., Lewis, E.A., Hansen, L.D., and Eatough, N.L., "The Chemical Composition of Environmental Tobacco Smoke III — Identification of Conservative Tracers of Environmental Tobacco Smoke," *Environ. Inter.,* 15:19–28 (1989).
18. Sinclair, J.D., Psota-Kelty, L.A., and Weschler, C.J., "Indoor/Outdoor Concentrations and Indoor Surface Accumulations of Ionic Substances," *Atmos. Environ.* 19:315–323 (1985).
19. Sinclair, J.D., Psota-Kelty, L.A., and Weschler, C.J., "Indoor/Outdoor Ratios and Indoor Surface Accumulations of Ionic Substances at Newark, NJ," *Atmos. Environ.,* 22:461–469 (1988).
20. Sinclair, J.D., Psota-Kelty, L.A., Weschler, C.J., and Shields, H.C., "Measurement and Modeling of Airborne Concentrations and Indoor Surface Accumulation Rates of Ionic Substances at Neenah, Wisconsin," *Atmos. Environ.,* 24:627–638 (1990).
21. Gunstone, F.D., Harwood, J.L., and Padley, F.B. (Eds.), *The Lipid Handbook,* 1st ed., (New York: Chapman and Hall, 1986).

CHAPTER 7

Methods for Chemical Characterization of Air Samples — PTEAM Prepilot Study

L. S. Sheldon, D. Whitaker, J. Sickles, E. Pellizzari, D. Westerdahl, and R. Wiener

INTRODUCTION

A nine-home chemical characterization study was initiated by the California Air Resources Board (ARB) and was performed as a collaborative effort with the U.S. Environmental Protection Agency's (EPA's) PTEAM Prepilot Study. The purpose of this study was to evaluate the sampling and analysis methods for identifying and quantitating chemical species in air samples. Four groups of chemicals were targeted for the study, namely polynuclear aromatic hydrocarbons (PAHs), phthalates, nitrosamines, and titratable acids. Indoor and outdoor fixed-site monitoring was performed for all chemical groups in nine homes for two 12-h periods. Testing during the nine-home study was designed to estimate precision, accuracy, and limits of detection for each method. In addition, results from field monitoring were examined to determine potential interferences resulting from the sample matrix, the percentage of samples where target analytes were above the detection limits, concentration levels of target analytes, and the relative distribution of target analytes between indoor and outdoor sample locations.

The EPA is undertaking a field monitoring study to estimate frequency distributions of human exposure to airborne particles of various size ranges. This particulate TEAM or PTEAM study, is being performed in two phases. Phase 1 is a nine-home study designed to field test monitoring and data collection activities associated with the proposed program. Phase 2 is a 200-home study designed to generate data for assessing exposure of the general population in a selected study

Table 1. Target Chemicals

Chemical Class		Compound
PAHs	3 ring	Acenaphthylene, phenanthrene, anthracene
	4 ring	Fluoranthene, pyrene, benzo(a)anthracene, chrysene
	5 ring	Benzo(k)fluoranthene, benzo(a)pyrene, Benzo(e)pyrene
	6 ring	Indeno(c,d)pyrene, benzo(g.h.i)perylene
	7 ring	Coronene
Nitrosamines		*N*-Dimethylnitrosamine *N*-Diethylnitrosamine *N*-Dipropylnitrosamine *N*-Dibutylnitrosamine *N*-Nitrosopiperidine *N*-Nitrosopyrrolidine *N*-Nitrosomorpholine
Acidity		Titratable acids, NO_3^-, SO_4^{-2}
Phthalates		Di-*n*-butylphthalate *n*-Butylbenzylphthalate Di-*n*-octylphthalate Di-2-ethylhexylphthalate

area to particles. Although the thrust of the PTEAM study is to evaluate particulate exposures, the target population for the Phase 2 study should also be a reasonable population for monitoring other pollutants. In this context, the ARB initiated this joint effort with EPA to use the PTEAM program as an efficient vehicle for monitoring chemicals in air inside and outside homes in California. This study was the ARB Phase 1 effort that was performed in conjunction with EPA's Phase 1 work.

The overall objective of the study was to evaluate sampling and analysis methods for monitoring selected chemical species in indoor and outdoor air samples. The selected approach was to perform a nine-home chemical characterization study using the proposed monitoring methods. Four groups of chemicals were targeted for the study: polynuclear aromatic hydrocarbons (PAHs), phthalates, nitrosamines, and titratable acids and related species. Specific pollutants are listed in Table 1. Indoor and outdoor fixed-site monitoring were performed for all chemical groups in all nine homes for two 12-h periods.

Materials and Methods

Selection of methods for the sampling and analysis of the target chemicals was an important component of the overall study. At the outset, several important decisions were made that affected the method selection process, including:

- Chemical characterization was to be performed only for indoor and outdoor fixed site samples. Personal exposure monitoring as proposed for the

PTEAM study was eliminated because of the high burden it would place on study participants.

- For more efficient field operations and to minimize the need for multiple pump evaluations, a single type of field sampler was to be used for all chemical classes. Sample collection rates of 5 to 10 l/min were used to provide sufficient material for analysis.
- Since the physical/chemical properties of the PAHs and phthalates are similar, these two chemical classes were to be collected together to minimize the sampling equipment required and the participant burden.
- For the organics, if possible, a particulate (<2.5 μm) and a vapor phase fraction were to be collected. The two fractions were to be combined prior to sample workup and analysis. Although this would not allow us to distinguish analyte concentrations in the particulate vs. vapor phase, it was intended to provide more sample material for a single analysis to increase the number of samples with analytical concentrations above the detection limit. It was also intended to reduce analytical costs.
- For acidity, only the particulate phase was to be collected and analyzed.

Given these constraints and since standard methods were not available, methods were proposed based on our own work with similar chemicals and reported literature methods,[1-4] our best judgement of the applicability of methods to our needs, the adaptability and simplicity of the method as used for field monitoring, and the overall method cost. Testing during this nine-home study was then intended to provide information in three important areas:

- Feasibility of field collection methods
- Performance of sampling and analysis procedures
- Approximate indoor and outdoor air concentrations of the chemicals of interest.

This paper will provide information on the analytical methods for the phthalates and the titratable acids. Method performance data and calculated sample concentrations will also be presented for these two groups of chemicals.

MATERIALS AND METHODS

Field Monitoring

A total of nine homes from the Glendora, Azusa, San Dimas, El Monte, Covina, and Baldwin Park areas of California were monitored during this study. These homes were purposely selected to provide a variety of aerosol levels, particulate size distributions, and emission sources in order to challenge the collection and analysis protocols.

Two locations at each house were monitored. One location was in the main living area inside the house and the other was an outdoor location near the most used entrance of the house. Sampling was performed during two consecutive 12-

Table 2. Samples Collected During Field Monitoring

Type	Number of Samples		
	Phthalates/PAHs	Nitrosamines	Titratable Acids
Indoor daytime	9	9	9
Indoor nighttime	9	9	9
Outdoor daytime	9	9	9
Outdoor nighttime	9	9	9
Field controls	4	4	8
Field blanks	4	4	4
Laboratory controls	4	4	8
Laboratory blanks	4	4	4
Duplicate samples	4	4	3

h time periods, providing overnight and daytime samples from each house. Each house was monitored for only 1 d; one house was sampled at a time. Field sampling was carried out from March 1 to March 23, 1989. The total number of samples collected is listed in Table 2.

All samples were collected using pumping systems that were designed and built by Battelle, Columbus Division. They utilized a Gast pump, a McMillan flow sensor, a Rustrak Ranger Data Logger, and a timer. Air flows were measured and recorded at the beginning and end of each 12-h sampling period using calibrated rotameters connected to the sampling heads with custom-made adapters.

Monitoring Methods for Phthalates

Particulate and vapor phase samples for phthalate analysis were collected using a quartz fiber filter backed by an XAD-resin bed. The nominal flow rate for sample collection was 15 l/min over the 12-h period to provide a sample volume of ~10 m^3. Collected samples were stored in the dark at room temperature in the field, then at –5°C once they were returned to the laboratory.

Phthalates were recovered from the XAD-2 resin and quartz fiber filters by soxhlet extraction with methylene chloride over a 16-h period. An extraction thimble was not used in order to reduce background contamination. The extracts were concentrated to 7 to 10 ml by Kuderna-Danish (K-D) evaporation. The methylene chloride was then solvent exchanged with hexane and further concentrated to 1.0 ml by nitrogen blowdown. Following concentration, the extracts were first analyzed for phthalates using a Hewlett Packard gas chromatograph with an electron capture detector. A 30 m DB-5 capillary GC column temperature programmed from 100 to 260°C at 5°C/min was used to separate the compounds.

Quantitation of target phthalates in sample extracts was performed using response factors calculated relative in the internal standard 2,3′, 4,4′, 6-pentachlorobiphenyl which was added at the same concentration to all standards and sample extracts. The concentration of phthalates in air samples was calculated by subtracting the average mass found in the field blanks from the mass found in each sample then dividing by the sample volume.

Monitoring Method for Titratable Acids

The titratable acids, nitrates, and sulfates were collected using the Battelle pumping system connected to an annular denuder-filter pack (DFP). The nominal flow rate through the denuder was 10 l/min over the 12-h sampling period for a sample volume of ~7 m^3. The DFP system consisted of a 120 mm annular denuder coated with citric acid and a 240 mm annular denuder coated with sodium carbonate. The filter pack contained a Teflon, a nylon, and a citric acid-coated cellulose filter. Using this method, all particles were collected on the Teflon filter. Prior to the filter collection, nitrate, sulfate, and ammonia (NH_3) gases were removed from the air stream using the sodium carbonate and citric acid denuder tubes, respectively. The Teflon filter was backed by a nylon filter and an acid-coated cellulose filter to collect nitric acid and ammonia gases that were either not collected by the denuders or vaporized off the particulate filter. Sample filters were stored in sealed plastic containers at 4°C until analyzed.

All three filters and the denuders were extracted by ultrasonication with deionized water. NO_3^- and SO_4^- were analyzed using ion chromatography (IC). The masses of NO_3^- and SO_4^- in each sample extract were calculated using response factors generated from standard solutions. Total mass of strong and weak acids in each extract was determined by Gran titration.[5] Only the extracts from the Teflon particulate filters were analyzed. The other extracts were saved and could have been analyzed if issues such as vapor phase concentration or filter collection efficiency required further investigation. The average mass of target chemicals in the field blanks was subtracted from the amount found in field samples. Sample concentrations were calculated by dividing the corrected mass by the sample volume as described above for the phthalates.

RESULTS AND DISCUSSION

Method Performance

The primary purpose of this pilot study was to evaluate the proposed analytical methods. Several types of quality control (QC) samples were used for this purpose. Field controls were intended to assess both accuracy and precision of the overall method. Field blanks were intended to provide information on background contamination and its variability. Field blank data were also used to estimate overall method detection limits and method quantitation limits. Laboratory blanks and controls and method blanks and controls were used as additional checks on accuracy, precision, and background contamination for the methods. Results from these latter QC samples were also used to determine sources of problems when the field controls or blanks showed unsatisfactory performance. Duplicate samples were intended to evaluate overall method precision for real samples.

Table 3. Performance Evaluation Data for Phthalate Analysis

Compound	Amount (ng) Found ± SD Field Blanks (N=4)	EMQL (ng/m³)	% Recovery ± SD Field Controls (n=4)[a]	Mean T RMD[b] for Duplicate Samples
Di-n-butyl-phthalate	385 ± 108	41	47 ± 23	NC[c]
n-Butylbenzyl-phthalate	47 ± 10	3.8	75 ± 16	4.5(1)[d]
Di-*n*-octyl-phthalate	9 ± 10	3.8	83 ± 19	25(2)
Di-2-ethylhexyl-phthalate	190 ± 68	26	69 ± 20[e]	17(2)

[a] All controls spiked with 1 mg of each target phthalate.
[b] Percent RMD = Percent relative mean deviation.
[c] Not found in any duplicate pairs.
[d] Number of duplicate pairs with measurable concentrations of target chemicals.
[e] N=3 — one very high value deleted.

Percent recovery for control samples was calculated as

$$\% \text{ Recovery} = \frac{M_C - M_B}{M_S} \times 100\%$$

where M_C is the mass of target found in the spiked controls, M_B is the mass of target found in the blanks, and M_S is the mass of target spiked into controls.

Estimated method quantitation limits (EMQL) were determined from the analysis of field blanks. These parameters were calculated as

$$\text{EMQL} = 3 \times \text{SD} / \text{sample volume}$$

where SD is the standard deviation of the amount of each phthalate found on field blanks. Although we had only four field blanks here, our approach is similar to that taken by the EPA where detection parameter estimates are based on the standard deviation of the calculated amount of analyte found in low level samples.

Performance evaluation data for the phthalates are shown in Table 3. Results presented here show generally acceptable precision and accuracy data. During laboratory setup, substantial effort was expended on reducing phthalate contamination that might result from sample processing. Analysis of both field and laboratory blanks showed trace levels of di-*n*-butyl- and di-*n*-octylphthalate (<50 ng per sample). Higher levels were found for butylbenzylphthalate and di-2-ethylhexylphthalate (~250 ng per sample), although these were well below the levels found on actual field samples as shown in Figure 1. The recovery of these latter two compounds was variable in control samples (percent relative mean deviation >20%) which was probably a result of sporadic background contamination. Alternate sample extraction procedures such as batch tumbling, ultrasonica-

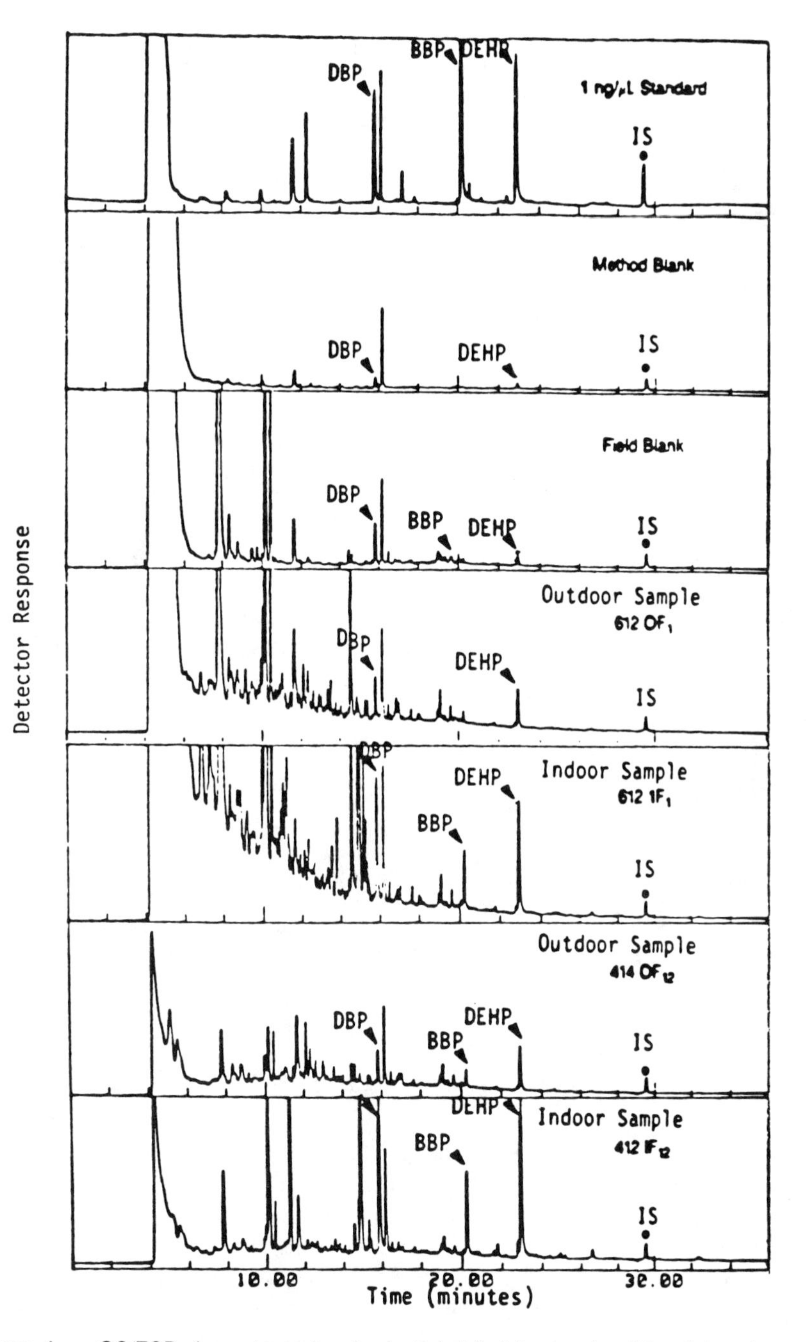

Figure 1. GC/ECD chromatographs of selected phthalate samples (IS — internal quantitation standard, DEHP — di-2-ethylhexylphthalate, BBP — butylbenzylphthalate, DBP — di-*n*-butylphthalate).

Table 4. Background NO_3^-, SO_4^{-2}, and Acid Levels on Blanks and Estimated Quantitation Limits (EMQL)

Compound	Field Blanks (N=3) amount/sample ± SD	EMQL (amount/m³)
NO_3^- (mg/sample)	1.06 ± 0.426[a]	0.32
SO_4^{-2} (μg/sample)	0.24 ± 0.45[a]	0.34
H^+ total (nmol/sample)	481 ± 132	99
H^+ strong (nmol/sample)	217 ± 63	47
H^+ weak (nmol/sample)	263 ± 79	59

[a] n=4.

Table 5. Percent Recovery for NO_3^-, SO_4^{-2}, and Titratable Acid in Field Controls

	Spike Level		% Recovery	
	High	Low	High (n=4)	Low (n=4)
NO_3^-	43.4 mg	4.34 mg	97 ± 5	97 ± 14
SO_4^{-2}	9.5 μg	.96 μg	98 ± 2.1	109 ± 29
Total H^+	900 n*M*	90 n*M*	126 ± 13	NC[a]
Strong H^+	200 n*M*	20 n*M*	360 ± 69	NC
Weak H^+	700 n*M*	70 n*M*	NC	NC

[a] Value not calculated, spike level below the EMQL.

Table 6. Precision of Duplicate Field Samples

Parameter	% Relative Mean Deviation
Total H^+	9.1 (2)[a]
NO_3^-	2.9 (3)
SO_4^{-2}	5.5 (3)

[a] Number of duplicate pairs with measurable values.

tion, or solvent elution of the resin bed are being considered to minimize contamination problems. Reproducibility of duplicate samples was considered acceptable.

Performance evaluation data for the titratable acids and related species are summarized in Tables 4 to 6. Results in these tables show excellent precision and accuracy for NO_3^- and SO_4^{-2} for field controls. Very little contamination was found on field blanks resulting in low detection limits (0.1 μg/m³). Results of duplicate samples also compared well (average percent relative mean deviation = 4.4).

Titratable acids were analyzed using a Gran titration that was intended to provide total H^+ concentrations plus weak and strong H^+ concentrations. Recovery

Table 7. Phthalate Concentrations (ng/m³) in Field Samples

	Di-*n*-butyl-phthalate		*n*-Butyl-benzyl-phthalate		Di-*n*-octyl-phthalate		Di-2-ethyl-hexylphthalate	
	In[a]	Out	In	Out	In	Out	In	Out
% Detected	47.4	5.3	100	26.3	57.9	26.3	89.0	47.3
Mean	110	20	47	5.5	15	3.0	81	24
D	160	19	39	12	29	3.7	38	23
Maximum	700	69	150	52	130	11	170	87
Minimum	ND[b]	ND	5.7	ND	ND	ND	ND	ND
Percentiles								
95	200	T	110	10	22	13	130	80
75	120	T	53	T	18	7.4	120	35
50	T	ND	31	ND	ND	ND	76	T
25	T	ND	16	ND	ND	ND	62	ND
5	ND	ND	7.0	ND	ND	ND	T	ND
n	17	17	17	17	17	17	17	17

[a] In = indoors; out = outdoors.
[b] T = less than the EMQL but greater than the EMDL. ND = less than the EMQL.

data for total H^+ was good (126 ± 13% for field controls). Unfortunately, both precision and accuracy for strong and weak H^+ were not acceptable and, therefore, quantitation could not be performed on these species. Background H^+ levels on field controls were high and variable, and resulted in a fairly high detection limit (33 nmol/m³). Good agreement was found for total H^+ analysis for replicate samples (percent relative mean deviations = 10%).

Sample Analysis Results

Major emphasis for this pilot study was placed on evaluating the proposed sampling and analysis methods, although field monitoring results were examined to gather preliminary information on pollutant levels. It should be stressed that this latter information on sample analysis results represents data from only a few purposely selected homes and should not be used to draw any conclusions or make any statements about phthalate or acidic pollutant levels in the study area or any other area in California.

Percent detected and summary statistics for phthalate concentrations calculated for indoor and outdoor air samples are given in Table 7. For all data analyses, concentrations found in duplicate pairs were averaged and used as a single value. Results show high percent values, although most phthalates were detected more frequently in indoor samples than in outdoor samples. A comparison of the summary statistics also shows that indoor concentrations were higher than outdoor concentrations in every case. Highest mean indoor concentrations were reported for di-*n*-butylphthalate (110 ng/m³) and di-2-ethylhexylphthalate (81 ng/m³). The highest maximum indoor concentration was reported for di-*n*-butylphthalate (700 ng/m³), while the highest median indoor concentration was calculated for di-2-ethylhexylphthalate (76 ng/m³).

Table 8. Comparison of Indoor/Outdoor Concentration Ratios for Phthalates

	Ratio of Indoor/Outdoor Concentration			
	Di-*n*-butyl-phthalate	*n*-Butyl-benzyl-phthalate	Di-*n*-octyl-phthalate	Di-2-ethyl-hexylphthalate
Median	3.3	26	0.84	4.3
Mean	16	46	15	7.2
SD	43	52	42	10
Maximum	175	155	180	40
Minimum	0.75	0.18	0.75	0.74
No. greater than 1	14	15	7	15
No. less than 1	2	1	9	1
n	16	16	16	16

Indoor/outdoor concentration ratios for samples collected at the same home during the same time period were calculated for all four phthalates. Results given in Table 8 again show that indoor phthalate concentrations were generally higher than outdoor concentrations with mean indoor/outdoor concentration ratios ranging from 7.2 for di-2-ethylhexylphthalate to 46 for *n*-butylbenzylphthalate. Median indoor/outdoor concentration ratios were greater than one for all four phthalates, although these concentration ratios were smaller than the mean concentration ratios. For di-2-ethylhexylphthalate and *n*-butylbenzylphthalate, all but one of the indoor samples had higher concentrations than the corresponding outdoor samples (i.e., the indoor/outdoor concentration ratio was greater than 1). For di-*n*-octylphthalate, 44% of the indoor samples had higher concentrations than the corresponding outdoor samples.

To further evaluate the data, statistics were calculated for samples collected indoors during the daytime and nighttime sampling periods. These data showed no trend toward either higher daytime or nighttime concentrations in indoor samples.

Percent detected and summary statistics for concentrations of titratable acids and related species are given in Table 9. NO_3^- and SO_4^{-2} ions were detected in all of the indoor and outdoor air samples. Titratable acid results are given only for total H^+ concentrations since analytical methods for distinguishing between the weak and strong acid species did not appear to be reliable.

Percent detected for total titratable acids were much lower than for nitrates and sulfates; 42% for outdoor samples and only 17% for indoor samples. The lower percent detected for total H^+ appears to be due to the fairly high EMQL that resulted from contamination in the field blanks.

A comparison of summary statistics shows higher outdoor concentrations for NO_3^- and titratable acids. For SO_4^{-2}, slightly higher levels were reported for indoor air samples. To further investigate the relationship between indoor and outdoor samples, indoor/outdoor concentration ratios were calculated for samples collected in the same home during the same time period. Results given in Table 10

Table 9. NO_3^-, SO_4^{-2}, and Titratable Acid (nm/m³) Levels in Field Samples

	Concentration					
	NO_3^- (mg/m³)		SO_4^{-2} (mg/m³)		Total Acid (nm/m³)	
	Indoor	Outdoor	Indoor	Outdoor	Indoor	Outdoor
% Detected	100	100	100	100	17	42
Mean	4.2	8.2	4.1	3.5	61	96
SD	4.7	5.9	2.8	1.9	65	52
Maximum	16.2	18.5	11.7	8.25	233	183
Minimum	0.47	0.77	0.85	0.65	ND	19.4
Percentile						
95	9.24	14.9	6.56	4.48	100	174
75	6.10	14	5.53	4.01	T[b]	109
50	2.01	5.93	3.16	3.75	T	T
25	0.62	3.40	2.28	1.69	ND[c]	T
5	0.52	2.07	2.01	1.29	ND	T
n	13	12	13	12	11	11

[a] Quantitated values greater than EMQL.
[b] Concentration less than EMQL but greater than EMDL.
[c] Concentration less than EMDL.

Table 10. Comparison of Indoor/Outdoor Ratios for NO_3^-, SO_4^{-2}, and Titratable Acids

	NO_3^-	SO_4^{-2}	Total H^+
Median	0.23	0.85	0.57
Mean[a]	0.51	1.48	0.71
SD	0.49	1.32	0.83
Maximum	1.7	4.4	3.0
Minimum	0.11	0.50	0
Number greater than 1	3	5	2
Number less than 1	10	8	10
n[b]	13	13	12

[a] Ratios where indoor levels less than the EMQL were assigned a value of 0. Ratios were both indoor and outdoor levels less than the EMQL were assigned a level of 1.
[b] Number of samples.

again show that NO_3^- and total H^+ concentrations are higher outdoors; whereas SO_4^{-2} levels are generally higher in indoor air samples.

The relationship between daytime and nighttime sample concentrations was also investigated, and although results are reported for only four homes, several trends are observed:

- NO_3^- species appear to have higher daytime concentrations in both indoor and outdoor samples

- SO_4^{-2} has slightly higher daytime concentrations in indoor samples while there appears to be no trend for outdoor samples
- Total H^+ may be slightly lower in daytime samples collected outdoors: no data is available for indoor samples because levels were all below the EMQL.

ACKNOWLEDGMENTS

Although the research described was funded by the California Air Resources Board (Contract No. A833-156) and the U.S. Environmental Protection Agency (Contract No. 68-02-4544), it has not been subjected to the required peer and administrative review and does not necessarily reflect the views of the Agencies, and no official endorsement should be inferred.

REFERENCES

1. Chuang, J., Hannan, S., and Wilson, N., *Environ. Sci. Technol.,* 21:798 (1987).
2. Sheldon, L. and Keever, J., "Collection and Analysis of Clean Room Air Samples," NASA Contractor Report 3947 (1985).
3. Sickles, J., Hodson, L., McClenny, W., Parr, R., Ellstad, T., Mulik, J., Anlauf, K., Webe, H., MacKay, G., Schiff, H., and Bubuc, D., *Atmos. Environ.,* 24:155 (1990).
4. Roundbehler, D.P., Reisch, J.W., Combs, J.R., and Fine, D.H., *Anal. Chem.,* 59:578 (1980).
5. MacDonald, T.I., Barker, B.J., and Caruso, J.A., *J. Clin.,* 49:202 (1972).

CHAPTER 8

Indoor Ozone Exposures Resulting from the Infiltration of Outdoor Ozone*

Charles J. Weschler, Datta V. Naik, and Helen C. Shields

ABSTRACT

Indoor and outdoor ozone concentrations were measured from late May through October at three office buildings with very different ventilation rates. The indoor values closely tracked the outdoor values, and, depending on the ventilation rate, were 20 to 80% of those outdoors. The indoor/outdoor data are adequately described with a mass balance model. The model can also be coupled with reported air exchange rates to estimate indoor/outdoor ratios for other structures. The results from this and previous studies indicate that indoor concentrations are frequently a significant fraction of outdoor values. These observations, and the fact that most people spend greater than 90% of their time indoors, indicate that indoor ozone exposure (concentration × time) is greater than outdoor exposure for many people. Relatively inexpensive strategies exist to reduce indoor ozone levels, and these could be implemented to reduce the public's total ozone exposure.

INTRODUCTION

More than half of the U.S. population resides in areas that have failed to meet the 120 ppbv National Ambient Air Quality Standard (NAAQS) for ozone.[1]

Ozone is recognized as causing acute and possibly chronic health effects.[1-7] Acute health effects in children have recently been reported at concentrations below 120 ppbv.[4,5] Results from the Harvard Six Cities Study, where measurements were obtained from school children, indicate that exposures to ozone at

* The following is from a technical paper that appeared in the December, 1989 issue of *JAPCA*, the journal of the Air & Waste Management Association.

Table 1. Measurements of Indoor/Outdoor Ozone Ratios

Structure	Location	I/O	Notes
Hospital	So. Calif.	0.67	UC Riverside group in 1971 (8)
Office/lab.	So. Calif.	0.80 ± 0.10	Shair's group (Cal Tech) in 1973;
Office/lab.		0.65 ± 0.10	home and natural ventilation (9)
Private home		0.70	
Hospital	So. Calif.	0.5	UC Riverside group in 1973; single
Swim. pool		0.5	day measurements; home used
2 Schools		0.3 – 0.7	evaporative cooler (10)
Office/lab.		0.5	
Private home		0.6	
2 Offices	So. Calif.	0.66	Shair's group (Cal Tech) in 1974;
2 Offices		0.54	max. and min. ventilation (11)
Office/lab.	So. Calif.	0.62	Shair's group (Cal Tech), 1974 (12)
5 Townhouses	Washington	0.5 – 0.7	GEOMET study conducted for EPA and
6 Apartments	Baltimore		HUD; 14-d monitoring at most sites;
2 Mobile hms.	Denver		reported in 1978 (13)
1 School	Chicago		
1 Hospital	Pittsburgh		
Prvt. homes	Medford, OR	0.1 – 0.25	LBL study of weatherized homes; reported in 1981 (14)
10 Prvt. hms.	Boston	0.2	GEOMET study conducted for EPRI;
2 Offices		0.3	reported in 1981 (15)
Art gallery	So. Calif.	0.5	Cass's group (Cal Tech) in 1983;
2 Museums		<0.1	museums used activated carbon air-filtration (16)
Art gallery	England	0.7 ± 0.1	U of East Anglia study; 1984 (17)
41 Prvt. hms.	Tucson	0.3	U of Arizona study; 1984 (18)
Museum	So. Calif.	0.45	Cass's group (Cal Tech) in 1986; 2-d period (19)
2 Museums (A)	So. Calif.	>0.67	Cass's group (Cal Tech) in 1989;
3 Museums (B)		0.30 – 0.40	museums (A) had high air ex. rates;
2 Museums (C)		0.10 – 0.20	museums (B) had conventional air
4 Museums (D)		<0.1	condt; museums (C) were tight bldgs; museums (D) had carbon air filt. (20)

* C.J. Weschler, D.V. Naik and H.C. Shields, Indoor Ozone Exposures, *JAPCA,* **39**, (1989).

concentrations below 78 ppbv are associated with transient decreases in lung function.[4] The New York University (NYU) field studies of children attending summer camp showed decrements in lung function even though the ozone concentration never exceeded 113 ppbv during the exposure period.[5] The NYU results suggest that the loss of lung function may be due to cumulative exposures to ozone rather than short term exposures to peak concentrations. Indeed, chronic effects may result from long-term or recurring exposures to ozone.[6,7] Given the results of studies on rats and monkeys, Lippmann believes that chronic exposure to ozone may cause accelerated aging of the human lung.[1]

Since the early 1970s investigators have reported indoor ozone concentrations that were a significant fraction (>0.5) of those measured outdoors. Table 1 summarizes results from these studies.[8-20] The earliest measurements were by Thompson's group at the University of California, Riverside and Shair's group at the California Institute of Technology. The Riverside group monitored indoor-outdoor levels for six air pollutants in 11 buildings in Southern California.[10] Measurements were made for only 1 d at each site. The indoor/outdoor (I/O) values for total oxidant (primarily ozone) ranged from 0.3 to 0.7. Shair's reported I/O values were also based primarily on short-duration measurements.[9,11,12] With the exception of a private residence, the monitored sites were buildings on the Cal Tech campus (ultimately 13 separate buildings). The majority of the reported I/O values were in the 0.5 to 0.8 range. In the late 1970s GEOMET conducted two major indoor studies[13,15] in which they monitored 7 to 15 pollutants, including ozone. In the first study, measurements were made in 15 structures in five different cities (Washington, Denver, Chicago, Baltimore, and Pittsburgh). I/O values for ozone were typically in the range of 0.5 to 0.7.[13] A subsequent study was conducted in 12 buildings in Boston; the I/O values averaged between 0.2 and 0.3.[15] Over the past decade, Cass's group at Cal Tech has examined ozone I/O values at numerous museums and art galleries.[16,19,20] In museums with high air exchange rates they reported I/O values approaching 0.7.[20] Despite the studies cited in Table 1, the misconception persists that ozone exposures (concentration × time) occur primarily outdoors and that indoor exposures can be ignored. An example comes from a "Policy Forum" that recently appeared in *Science*:[21] "There is mounting evidence of chronic effects from longer term or recurring exposures (to ozone) at or below levels of acute concern. Human exposure, though, is *limited to time out of doors because structures protect occupants*" (emphasis added).

In this paper we present additional evidence that indoor ozone concentrations can be a significant fraction of outdoor concentrations. Simultaneous indoor and outdoor measurements, with only minor interruptions, have been made for a complete ozone season in New Jersey (150 d). The results can be described with a mass-balance model, and this model can be used to estimate indoor/outdoor ratios in other structures across the country. Based on such estimates, we conclude that indoor ozone *exposures* are frequently greater than outdoor *exposures*. It is important that public health officials and policy makers be aware of indoor exposures, so that they can properly assess health risks and prioritize strategies to reduce the individuals' total ozone exposure. Indeed, measures do exist to reduce indoor ozone concentrations.[16,22,23]

EXPERIMENTAL

During the 1988 summer ozone season, extensive simultaneous indoor and outdoor ozone measurements were made at an office-laboratory complex in suburban New Jersey, approximately 27 air-miles from Manhattan and 6.5 air-miles west of the Atlantic Ocean (Figure 1). The complex, constructed during 1983 and

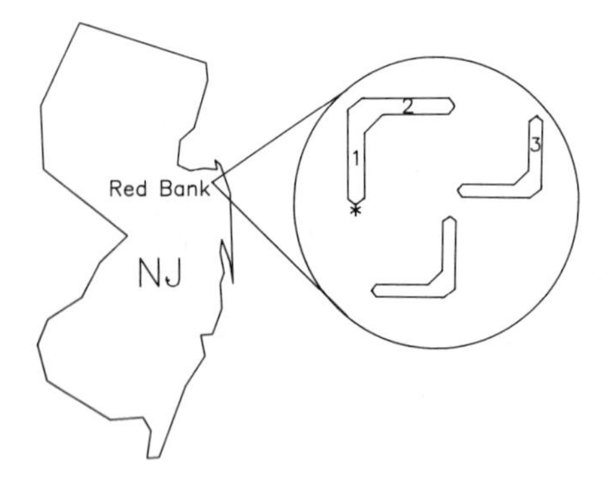

Figure 1. Location of office complex where ozone measurements were made. The insert shows the three major buildings at the complex; arabic numerals designate the indoor sampling locations: Sites 1, 2 and 3; the "*" designates the outdoor sampling location.

1984, consists of three large buildings and one smaller structure. The indoor measurements occurred at three different locations: on the second floor of the south side of Building 3 (hereafter referred to as Site 1), on the second floor of the east side of Building 3 (hereafter referred to as Site 2), and on the fourth floor of the north side of Building 1 (hereafter referred to as Site 3). Building 3 is a three-story structure with approximately 425 occupants. It contains two compartments separated by an atrium. Building 1 is a four-story structure with approximately 530 occupants that also has two separated compartments. Sites 1 and 2 each have a floor area of 3257 m^2 and a total volume of 8940 m^3; Site 3 has a floor area of 1424 m^2 and a volume of 3910 m^3. Each floor on each side of these buildings has its own constant volume air handling system. For each of the indoor sites examined in this study, the ventilation rate is assumed to equal the sum of the exhaust rates. At each of the sites the exhaust rates are fixed; the total exhaust rates are 9.3 m^3/s for Site 1, 20.3 m^3/s for Site 2, and 0.68 m^3/s for Site 3. These are not simply the rated values of the various exhaust fans, but are actual measured exhaust rates for each of the fans. Given the volumes stated above, these exhaust rates result in air exchange rates of 4.0 air changes/hr (ach) at Site 1, 8.2 ach at Site 2, and 0.6 ach at Site 3.

Ozone concentrations were measured with two UV photometric ozone analyzers (Dasibi Model 1003 AH; wavelength: 254 nm; range: 0 to 500 ppbv; sensitivity: 1 ppbv; accuracy: 3%; precision: ±1% or 1 ppbv, whichever is greater). The air was sampled at a rate of 2 l/min using Teflon® tubing; an inline 0.5 micron Teflon® filter was used to remove airborne particles. The filter was replaced every 2 weeks, and instrument performance checks were conducted weekly.

The data, collected at 30-sec intervals, were stored on floppy disks by interfacing these instruments to personal computers. Outdoor ozone readings were recorded continuously from May 31 to October 28, 1988. From May 31 to August

11 indoor ozone readings were obtained at Site 1. For the rest of the period, the indoor monitor was rotated weekly among the three indoor sites.

RESULTS

Two representative weeks of indoor (Site 1)-outdoor ozone data are presented in Figure 2. The indoor values are typically 40 to 60% of the outdoor values. Furthermore, the indoor concentrations closely track the outdoor concentrations. Even relatively quick, small outdoor changes are mirrored indoors. See, for example, the afternoon of June 20, 1988, Figure 2A, or the afternoon of July 9, 1988, Figure 2B. This characteristic was observed for all of the indoor-outdoor data collected (150 d of almost continuous measurements).

Indoor-outdoor ozone measurements for 1 week sampling intervals at each of the three indoor sites are contrasted in Figure 3. The indoor/outdoor ozone ratios are largest for the site with the highest ventilation rate (Site 2, Figure 3A) and lowest for the site with the lowest ventilation rate (Site 3, Figure 3B). Site 1 (Figure 3C) falls between these extremes. During periods when the outdoor ozone concentration was relatively constant, the median values for the indoor/outdoor ratios were 0.71 for Site 2, 0.22 for Site 3, and 0.54 for Site 1.

The number of days in 1988 that the indoor ozone concentrations exceeded various benchmark values at Sites 1 to 3 are listed in Table 2. At Site 2 there were 4 d when the indoor concentration exceeded 120 ppbv and 17 d when it exceeded 80 ppbv, while at Site 1 there were 7 d when the indoor values were greater than 80 ppbv. It is instructive to consider indoor ozone concentrations in excess of benchmark values in various urban areas, given that a certain number of buildings in each city will have I/O values similar to Sites 1 and 2. In 1988 the second highest ozone concentrations measured outdoors in New York, Philadelphia, Chicago, Houston and Los Angeles were 217, 200, 215, 220 and 330 ppbv, respectively.[24] On the cited days in these cities, buildings with ozone I/O values greater than 0.6 had *indoor* concentrations that were greater than 120 ppbv. Indeed, in Los Angeles an I/O value of only 0.36 was sufficient to yield an indoor ozone concentration greater than 120 ppbv. I/O values of 0.4 to 0.6 are not unusual; see, for example, Table 1 and the discussion that follows.

Since the outdoor ozone levels strongly influence the indoor ozone levels, several features of the outdoor data are worth noting. The outdoor concentrations display the expected diurnal variations. On cloudless days the ozone levels tend to be highest in the afternoon (2:00 to 4:00 pm) and lowest near dawn (6:00 to 8:00 am). However, the influence of transported pollution plumes is also suggested. Ozone concentrations were observed to peak at multiple periods throughout the day and night. Indeed, peaks would frequently be observed after sunset (e.g., approximately 100 ppbv from 10:00 pm to midnight on June 22, 1988, Figure 2A; approximately 70 ppbv from 2:00 to 4:00 am on July 8, 1988, Figure 2B; approximately 90 ppbv from 3:00 to 5:00 am on July 14, 1988, not shown). These

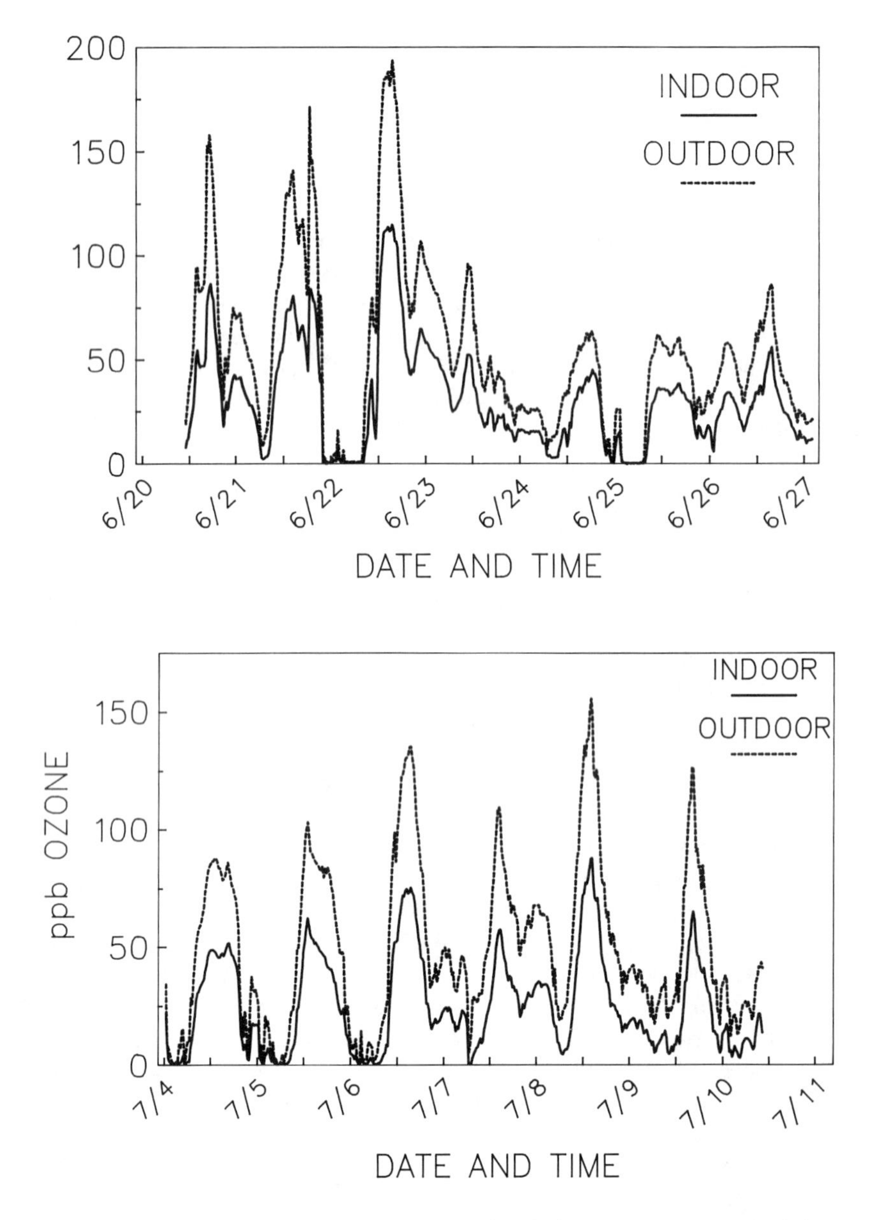

Figure 2. Representative indoor (—) and outdoor (---) ozone measurements during the summer of 1988 at an office complex in Red Bank, NJ. Indoor values are for Site 1. Tick marks indicate beginning and noon of the stated day. Note the ozone plumes after sunset on (top) 6/21, 6/22–6/23, 6/25, 6/26 and (bottom) 7/8.

observations are consistent with earlier reports on the movement and evolution of photochemical pollutants from urban areas across regions situated downwind.[25-27]

As a consequence of transported pollution plumes, ozone concentrations frequently remain elevated for 8 h or more. This is evident in Figures 2 and 3. On some occasions (e.g., August 17–18, 1988, Figure 3A) ozone concentrations were elevated throughout a 24 h period. It follows that in such locations effective exposure is not limited to a few hours at midday, but extends over the entire pollution episode.

DISCUSSION

Sources of Indoor Ozone

The fact that the indoor and outdoor ozone levels closely track (see Figures 2 and 3) indicates outdoor ozone is primarily responsible for changes in the indoor level. One might argue that indoor sources contribute a constant background upon which these changes are superimposed. Office equipment that uses high voltage or ultraviolet light can generate ozone, and, consequently, certain photocopiers and laser printers are potential indoor sources.[28] Each of the three sites contains two or three photocopiers and about a dozen laser printers. Nonetheless, the data indicates that *most of the indoor ozone originates outdoors*. This is apparent from the low indoor ozone concentrations during work hours on those days with low outdoor ozone concentrations. For example, indoor ozone levels less than 10 ppbv have been recorded at 10:00 am on August 16, 1988 (Figure 3A), at noon on August 25, 1988 (Figure 3B), and at 10:00 am on August 31, 1988 (Figure 3C). Consequently, even if indoor sources were contributing a constant background, the background could be no larger than 10 ppbv.

Mass-Balance Model

The indoor-outdoor data collected at this building complex throughout the ozone season can be explained in the context of a mass balance model.[11,29] Assuming that the outdoor ozone concentration is not changing rapidly, that indoor sources make an insignificant contribution to the steady-state indoor ozone concentration, and that building filters and leakage paths remove a negligible quantity of ozone, then:

$$\frac{I}{O} = \frac{(v_1 + v * f)}{k_d A_a + (v_1 + v * f)} \tag{1}$$

where

I/O = the ratio of the indoor ozone concentration to the outdoor ozone concentration

v_1 = the volume of air leaking into or out of the building per unit time (m^3/s)

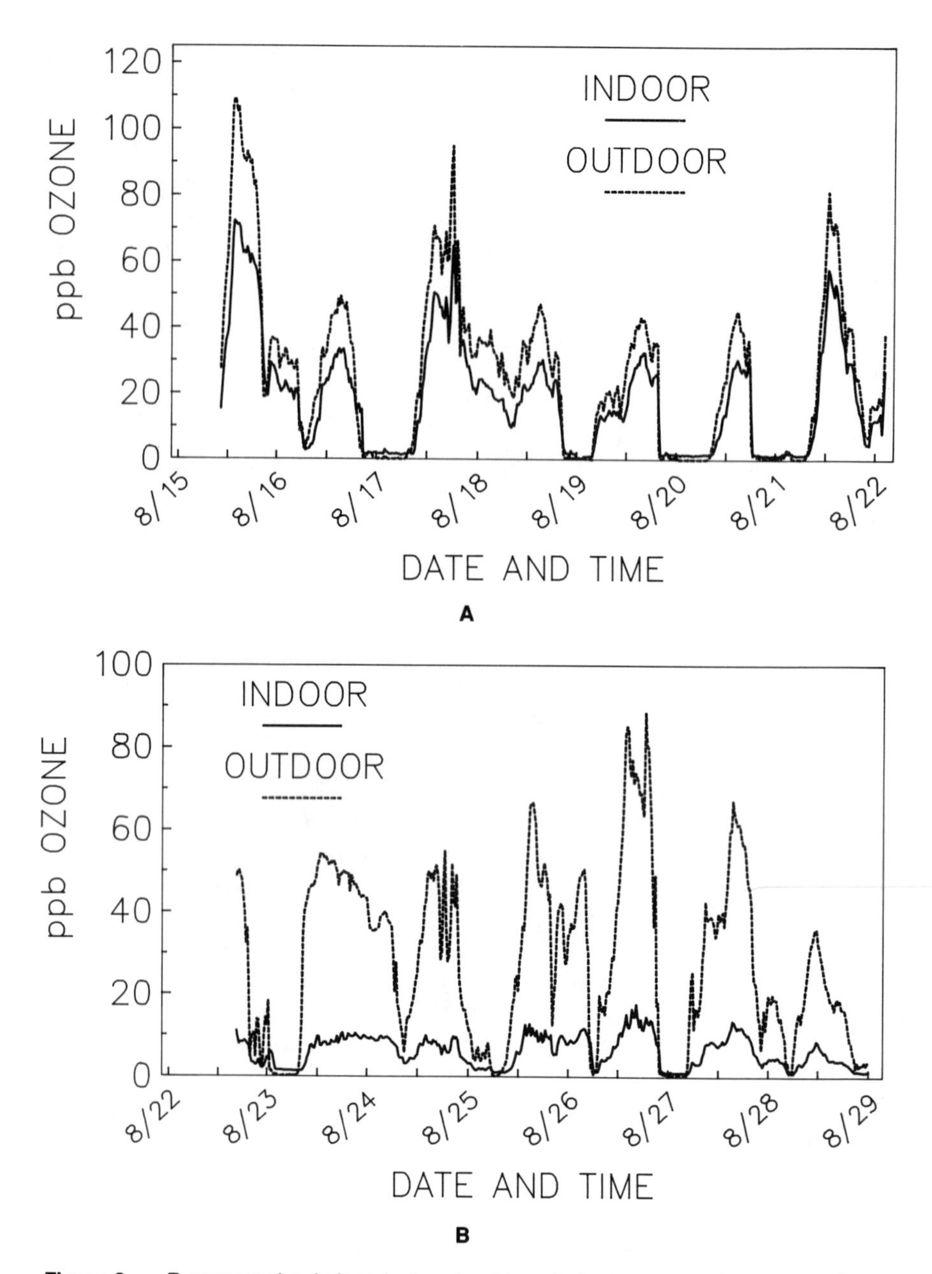

Figure 3. Representative indoor (—) and outdoor (---) ozone measurements at three different indoor sites from 8/15 through 9/6/88. Each site has a very different ventilation rate, and, consequently, a different indoor/outdoor ratio (I/O) during steady-state conditions: (A) Site 2, 8.2 air changes/h (ach), median I/O = 0.71; (B) Site 3, 0.6 ach, median I/O = 0.22; and (C) Site 1, 4.0 ach, median I/O = 0.54.

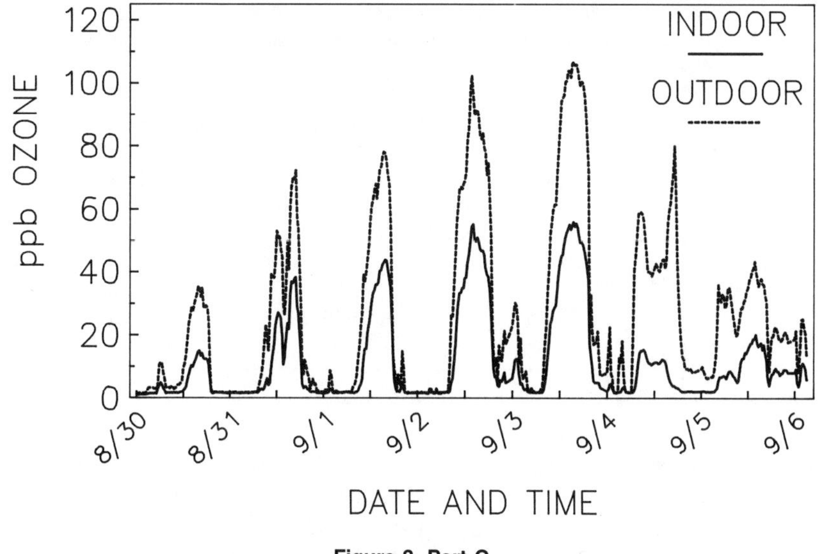

Figure 3. Part C.

v^* = the air flow in the air handling system (m^3sec^{-1})
f = the fraction of circulation made up with outside air
k_d = the ozone deposition velocity (m sec^{-1})
A_d = the internal ozone deposition area (m^2)

This expression simply states that for a unit of time when steady-state conditions apply, the indoor/outdoor ratio of ozone is equal to the total amount of outside air that enters the building divided by the sum of ozone scavenged by interior surfaces and the total amount of air that exits the building.

For the three indoor sites examined in this study, ($v_I + v^*f$) is assumed to equal the total exhaust rates. A_d is estimated as eight times the floor area, following the example of Mueller et al.[23] The respective ($v_I + v^*f$) and A_d values are: 9.9 m^3/s, 26,000 m^2 for Site 1; 20.3 m^3/s, 26,000 m^2 for Site 2; and 0.68 m^3/s, 11,400 m^2 for Site 3. The exhaust rate for Site 3 is much smaller than those for the other two sites because there are no laboratory fume hoods in Site 3. The ozone deposition velocity, k_d, used in Equation 1 actually represents a weighted average of the rate at which ozone reacts with different surfaces in the building. Such rates have been reported in several studies.[9,19,23] Shair[22] calculated an average value of 3.6×10^{-4} m/s for k_d, based on his measurements in laboratory-office buildings. We will use the same value. (The physical significance of k_d is clarified by stating it as: [unit volume/unit area] per unit time, e.g., $m^3/m^2/s$. Hence, it expresses the rate at which a unit volume of ozone is scavenged by a unit area of a given surface; it is a normalized flux, not a physical velocity. Ozone decay on surfaces follows first order kinetics. A deposition velocity of 3.6×10^{-4} m/s is equivalent to a 1st order

Table 2. Number of Days During the Summer of 1988 That Indoor Ozone Concentrations Exceeded the Listed Values at an Office Complex in Red Bank, NJ

Ozone Conc. (ppbv)	Days exceeding the listed ozone concentration		
	Site 1 (I/O=0.54)	Site 2 (I/O=0.71)	Site 3[a] (I/O=0.22)
120	0	4	0
100	1	10	0
80	7	17	0
60	18	40	0
40	52	69	1

[a] The three indoor sites have different ventillation rates.

equivalent to a 1st order rate constant of $1 \times 10^{-3}\ s^{-1}$ ($t_{1/2} = 11.1$ min), assuming that these office buildings have a surface to volume ratio of 2.9 m^{-1}.)

Substituting the above values in Equation 1 yields indoor/outdoor ratios for ozone of 0.51 at Site 1, 0.68 at Site 2, and 0.14 at Site 3. These calculated indoor/outdoor ratios are in good agreement with the median values (0.54, 0.71, and 0.22 for Sites 1, 2, and 3, respectively) obtained during periods when the outdoor ozone concentration was relatively constant (see, for example, Figure 3).

It should be noted that the present study is not unique in applying a model to relate indoor ozone concentrations to those outdoors. A number of earlier studies have also developed models to predict or explain indoor ozone levels.[11,19,30] The use of the steady-state approximation in the present work does make our model somewhat simpler to apply than those employed in certain earlier studies.

Equation 1 can be reformulated in terms of air exchange rates if its numerator and denominator are divided by the volume (Vol) of the indoor site:

$$\frac{I}{O} = \frac{E_x}{k_a(A_a / Vol) + E_x} \tag{2}$$

where

E_x = $(v_1 + v^*f)/Vol$ = the air exchange rate, typically expressed in units of air changes per hour (ach), and

(A_d/Vol) = the surface to volume ratio for the structure (m^2/m^3); a reasonable value is 2.9 m^{-1}, derived from the studies of Mueller et al.[23]

Equation 2 has the advantage of permitting rapid estimates of I/O ratios for ozone based on measured or codified air exchange rates. Indeed, where the appropriate input parameters exist, Equation 2 adequately describes the indoor-outdoor data from the earlier studies cited in Table 1. This further supports the use of Equation 2 for the estimation of I/O ratios in other structures, both commercial and residential. The utility of this formulation is demonstrated in the discussion that follows.

Indoor Ozone Exposures

Time activity studies show that people spend greater than 90% of their time indoors.[31] For people who are very young, very old, or ill, the percentage is even higher. Because of the large amount of time spent indoors, some indoor exposures can be greater than the outdoor exposures, even when the indoor concentrations are less than those outdoors. This is frequently the case for ozone. Several examples illustrate this point.

A recent survey of 600 California homes found a median air exchange rate of 1.5 ach during the month of July.[32] Substituting this value in Equation 2, together with a surface-to-volume ratio of 2.9 m^{-1} and a deposition velocity of 3.6×10^{-4} m/s, yields an indoor/outdoor ratio of 0.28. Consider now the ozone exposure for people who spend their indoor time in homes where the ozone concentration is 28% of the outdoor value. We will make some very conservative assumptions: (a) indoor nighttime ozone concentrations are zero (they frequently are not — see Figures 2 and 3); (b) all of one's time outdoors occurs during daylight hours; (c) there are only 13 h of significant outdoor ozone levels per summer day of which the average person spends 10.6 h indoors and 2.4 h outdoors (on average, it will be less); and (d) the average outdoor ozone concentration is X ppbv during these 13 h. Then outdoor ozone exposure is (2.4 hrs) × (X ppbv) = 2.4 X ppbv h, and indoor ozone exposure is (10.6 h) × (0.28 X ppbv) = 3.0 X ppbv h. Hence, in this Southern California scenario, the indoor ozone exposures are conservatively estimated to be 1.25 times greater than the outdoor ozone exposures.

A second example comes from a recent EPA study of 45 homes in Los Angeles.[33] During the month of July, investigators measured median daytime air exchange rate of 2.2 to 2.4 ach. (Most of these homes did not have air conditioning.) At this air exchange rate, the indoor ozone concentration is about 38% of the outdoor value. Using the same conservative assumptions employed in the first example, the indoor exposures for people living in these homes are 1.7 times greater than their outdoor exposures.

Actually, in homes with natural ventilation the indoor ozone concentrations, and consequently the indoor ozone exposures, can be much larger than in these examples. The California Department of Health Services has examined the effects of window-opening on residential ventilation.[34] They found that in some cases opening windows increased the ventilation rate by as much as eleven times the original rate. With windows open, the air exchange rate can exceed 5 ach. Frequently people open their windows and turn on fans as the outdoor temperature increases. Ozone concentrations also tend to increase with the outdoor temperature. Hence, in homes without air conditioning, the ventilation rate may be at its highest value when the ozone concentrations are also at their highest value. In such situations, the indoor ozone exposures may easily be three to four times the outdoor ozone exposures.

The situation in commercial buildings is somewhat different. Most commercial buildings have mechanical air handling systems. Some of these systems have "fixed" air exchange rates while other systems can be adjusted as dictated by prevailing weather conditions and occupant needs. In the latter case, building operators tend to reduce ventilation rates as the outdoor temperature rises in order to conserve "cooled air" (this saves energy, although it can sometimes lead to elevated concentrations of pollutants with indoor sources.)[35,36] Hence the air exchange rates tend to be lower, resulting in lower I/O values, when the outdoor ozone levels are higher. Reduced ventilation rates are less likely to be in effect when ozone plumes come through a region after the peak temperatures of the day. This is most likely to occur in regions downwind of urban areas that generate large quantities of ozone precursors.

It is instructive to examine air exchange rates that have actually been measured in commercial facilities. Persily and coworkers at the National Institute of Standards and Technology (formerly NBS) have made careful measurements of air exchange rates in 14 commercial buildings for extended periods of time.[37] The average air exchange rate for these 14 structures is 0.94 ach. The mean air exchange rates measured in individual buildings ranged from 1.73 to 0.29 ach. In 1988 the US EPA concluded a detailed study of indoor air quality in public buildings. As part of this work, they measured air exchange rates in a home for the elderly, a school, and an office. The median air exchange rates were 1.7, 0.9, and 0.8 ach, respectively.[38] To put these numbers in perspective, for the assumptions used above, an air exchange rate greater than 1.1 ach will result in an indoor exposure greater than the outdoor exposure.

For commercial buildings with mechanical ventilation systems, air exchange rates can be estimated from building codes and standard practices. Air exchange rates derived in this fashion from the American Society of Heating, Refrigerating and Air Conditioning Engineers (ASHRAE) outdoor air requirements for commercial facilities are listed in Table 3.[35,36] The occupant densities listed in the table were calculated using the occupancies in the ASHRAE standard (persons per 100 m^2 floor area) and assuming a height of 3 m for floor-to-ceiling plus concealed space. Several of the estimated air exchange rates for different types of indoor space are worth noting. In office space, the revised ASHRAE standard recommends a minimum of 0.010 m^3 outside air/s per person. For the listed occupant density, this results in a minimum air exchange rate of 0.84 ach; the corresponding I/O ozone ratio calculated from Equation 2 is 0.18. In classrooms, where the occupant density is much larger, the air exchange rate derived from the ASHRAE standard is 4.5 ach. At such an air exchange rate, the indoor ozone concentration is calculated to be 0.54, more than half that outdoors. Similarly high air exchange rates and I/O ozone ratios are estimated for any space where the occupant density is large (e.g., conference rooms, dining areas, auditoriums, theatres, transportation waiting rooms, etc.). The fractional exposure that each of these "microenvironments" contributes to a person's total ozone exposure depends on the amount of time that a person spends in each location. A high I/O for ozone in a classroom,

Table 3. Air Exchange Rates Derived from the American Society of Heating, Refrigerating, and Air-Conditioning Engineers' (ASHRAE's) Outdoor Air Requirements for Commercial Facilities[a]

Minimum outside air requirement (m^3/s per person)	Estimated max. occupancy (persons per 300 m^3)	Air exchange rate (ach)	Calc. I/O	Application
0.0025	7	0.21	0.05	Office space, previous ASHRAE std. (35)
0.010	7	0.84	0.18	Office space, revised ASHRAE std. (36)
0.0125	10	1.5	0.28	Patient rms, in hospitals & nursing hms., rev ASHRAE (36)
0.0075	50	4.5	0.54	Classrooms, revised ASHRAE std. (36)
0.010	50	6.0	0.61	Hotel conference rooms, revised ASHRAE std. (36)
0.010	70	8.4	0.69	Dining areas in restaurants, revised ASHRAE std. (36)
0.030	70	25.2	0.87	Smoking lounges in public areas, rev ASHRAE std. (36)

[a] Occupancies (persons per 300 m^3) were calculated from the suggested number of occupants per 100 m^2 floor area, assuming a 3-m height for floor to ceiling plus concealed space. The corresponding indoor/outdoor (I/O) values for ozone, calculated using Equation 2, are also listed.

occupied for 6 h or longer, is potentially more significant than the same ratio in a theater or restaurant, occupied for shorter periods. However, neither situation has to be tolerated. As discussed in the next section, high indoor ozone concentrations can be prevented.

Methods to Reduce Indoor Ozone

Researchers have demonstrated that appropriate filters can significantly reduce indoor ozone concentrations in buildings with mechanical ventilation systems. In a 3-year study conducted in Southern California, Shair measured the efficacy of an auxiliary filtration unit consisting of activated charcoal filters.[22] The fraction of ozone removed by the filters was 0.95, 0.8, and 0.5 at the end of the first, second, and third years, respectively. The yearly increase in operational costs was relatively small ($600 in 1975 dollars for a building with approximately 100 occupants), while the capital investment was about $12,000. Studies by Mueller et al.[23] and Shaver et al.[16] have also shown that activated carbon can be effective in removing ozone from outdoor air. Indeed, several of the very low I/O values in Table 1 are for buildings equipped with carbon filtration.[16,20] Noble metal catalysts, similar to those used by the aircraft industry,[39] might also be used to reduce ozone concentrations in buildings with mechanical air handling systems. In buildings with natural ventilation, occupants can be encouraged to open their windows during the cooler part of the day, typically late evening and early morning, and to close their windows and retain the cool air during the warmest part of the day. This will reduce ventilation rates during the period when ozone concentrations are frequently highest. However, any reduction in ventilation rates should be approached with caution. A number of studies have demonstrated that pollutants with indoor sources, such as certain volatile organic compounds, can accumulate to undesirable concentrations when building ventilation is inadequate.[35,36,38,40]

CONCLUSIONS

This paper demonstrates that, for many structures, indoor ozone concentrations can be a significant fraction of outdoor levels. Because of the large amount of time that people spend indoors, indoor ozone exposures are frequently greater than outdoor exposures. Individuals, as well as public officials, should be cognizant of this fact to properly assess health risks and to craft policies for reducing ozone exposures. Since most indoor ozone originates outdoors, efforts to reduce outdoor ozone concentrations are essential. If outdoor levels are reduced, indoor levels will also be reduced. While ventilation of both private and public buildings is extremely important to prevent accumulation of pollutants from indoor sources,[35,36,39,40] an undesired result of ventilation can be significant indoor ozone concentrations. As noted above, strategies exist to reduce indoor ozone levels. Such measures are especially recommended for structures such as hospitals, nursing homes, and schools. However, mitigation strategies will not be implemented if the misconception persists that buildings adequately protect occupants from ozone.

ACKNOWLEDGMENTS

Discussions with Paul J. Lioy of the University of Medicine and Dentistry of New Jersey were extremely helpful. We thank Leonilda A. Farrow of Bellcore for assistance in interfacing the personal computers to the ozone meters. DVN gratefully acknowledges support from the Bellcore-Monmouth College Cooperative Summer Faculty Research Program.

REFERENCES

1. Lippmann, M., "Health effects of ozone, a critical review," *JAPCA,* 39:672 (1989).
2. "Air Quality Criteria for Ozone and Other Photochemical Oxidants, Vols. I–V," EPA/600/8-84/020aF, ECAO, Research Triangle Park, NC (August 1986).
3. Marshall, E., "Clean air? Don't hold your breath," *Science,* 244:517 (1989).
4. Kinney, P.L., Ware, J.H., Spengler, J.D., Dockery, D.W., Speizer, F.E., and Ferris, Jr., B.G., "Short-term pulmonary function change in association with ozone levels," *Am. Rev. of Resp. Dis.,* 139:56 (1989).
5. Spektor, D.M., Lippmann, M., Lioy, P.J., Thurston, G.D., Citak, K., James, D.J., Bock, N., Speizer, F.E., and Hayes, C., "Effects of ambient ozone on respiratory function in active normal children," *Am. Rev. of Resp. Dis.,* 137:313 (1988).
6. Huang, Y., Chang, L.Y., Miller, F.J., Graham, J.A., Ospital, J.J., and Crapo, J.D., "Lung injury caused by ambient levels of oxidant air pollutants: extrapolation from animal to man," *Am. J. Aerosol Med.,* 1:180 (1988).
7. *Air Quality Guidelines for Europe,* Vol. 23, (Copenhagen: World Health Organization Regional Publications, 1987), pp. 318–323.
8. Thompson, C.R., "Measurement of total oxidant levels at Riverside Community Hospital," *Arch. Environ. Health,* 22:514 (1971).
9. Sabersky, R.H., Sinema, D.A., and Shair, F.H., "Concentrations, decay rates, and removal of ozone and their relation to establishing clean indoor air," *Environ. Sci. Technol.,* 7:347 (1973).
10. Thompson, C.R., Hensel, E.G., and Kats, G., "Outdoor-indoor levels of six air pollutants," *JAPCA,* 23:881 (1973).
11. Shair, F.H. and Heitner, K.L., "Theoretical model for relating indoor pollutant concentrations to those outside," *Environ. Sci. Technol.,* 8:444 (1974).
12. Hales, C.H., Rollinson, A.M., and Shair, F.H., "Experimental verification of linear combination model for relating indoor-outdoor pollutant concentrations," *Environ. Sci. Technol.,* 8:452 (1974).
13. Moschandreas, D.J., Stark, J.W.C., McFadden, J.E., and Morse, S.S., "Indoor air pollution in the residential environment," Volumes I and II. Final Report. GEOMET EF-668, Contract No. 68-02-2294, U.S. Environmental Protection Agency, Environmental Research Center, and U.S. Department of Housing and Urban Development, Office of Policy Development and Research, EPA 600/7-78/229A (1978).
14. Berk, J.V., Young, R.A., Brown, S.R., and Hollowell, C.D., "Impact of energy-conserving retrofits on indoor air quality in residential housing," LBL-12189. Lawrence Berkeley Laboratory, Jan. 1981. (Presented at 74th Annual Meeting of the Air Pollution Control Association, Philadelphia, June 1981).
15. Moschandreas, D.J., Zabransky, J., and Pelton, D.J., "Comparison of indoor and outdoor air quality," Electric Power Research Institute, Report EA-1733, Research Project 1309, Palo Alto, CA (March 1981).
16. Shaver, C.L., Cass, G.R., and Druzik, J.R., "Ozone and the deterioration of works of art," *Environ. Sci. Technol.,* 17:748 (1983).
17. Davies, T.D., Ramer, B., Kaspyzok, G., and Delany, A.C., "Indoor/outdoor ozone concentrations at a contemporary art gallery," *JAPCA,* 34:135 (1984).
18. Lebowitz, M.D., Corman, G., O'Rourke, M.K., and Holberg, C.J., "Indoor-outdoor

air pollution, allergen and meteorological monitoring in an arid southwest area," *JAPCA,* 34:1035 (1984).

19. Nazaroff, W.M. and Cass, G.R., "Mathematical modeling of chemically reactive pollutants in indoor air," *Environ. Sci. Technol.,* 20:924 (1986).
20. Druzik, J.R., Adams, M.S., Tiller, C., and Cass, G.R., "The measurement and model predictions of indoor ozone concentrations in museums," *Atmos. Environ.,* in press.
21. Russell, M., "Ozone pollution: the hard choices," *Science,* 240:1275 (1988).
22. Shair, F., "Relating indoor pollutant concentrations of ozone and sulfur dioxide to those outside: economic reduction of indoor ozone through selective filtration of the make-up air," *ASHRAE Transactions,* 87(I):116 (1981).
23. Mueller, F.X., Loeb, L., and Mapes, W.H., "Decomposition rates of ozone in living areas," *Environ. Sci. Technol.,* 7:342 (1973).
24. U.S. Environmental Protection Agency, "A comparison of 1988 ozone concentrations to 1983 and 1987 ozone concentrations," Monitoring and Reports Branch, Technical Support Division, Office of Air Quality Planning and Standards, Research Triangle Park, NC, (July 27, 1989).
25. Cleveland, W.S. and Graedel, T.E., "Photochemical air pollution in the northeast United States," *Science,* 204:1273 (1979).
26. Cleveland, W.S., Kleiner, B., McRae, J.E., and Warner, J.L., "Photochemical air pollution:transport from the New York City area into Connecticut and Massachusetts," *Science,* 191:179 (1976).
27. Cleveland, W.S. and Kleiner, B., "Transport of photochemical air pollution from Camden-Philadelphia urban complex," *Environ. Sci. Technol.,* 9:869 (1975).
28. Allen, R.J., Wadden, R.A., and Ross, E.D., "Characterization of potential indoor sources of ozone," *Am. Ind. Hyg. Assoc. J.,* 39:466 (1978).
29. Weschler, C.J., Shields, H.C., Kelty, S.P., Psota-Kelty, L.A., and Sinclair, J.D., "Comparison of effects of ventilation, filtration, and outdoor air on indoor air at telephone office buildings: a case study," in *Design and Protocol for Monitoring Indoor Air Quality, ASTM STP 1002,* N.L. Nagda, and J.P. Harper, (Eds.), (Philadelphia, 1989), pp. 9-34.
30. Allen, R.J. and Wadden, R.A., "Analysis of indoor concentrations of carbon monoxide and ozone in an urban hospital," *Environmental Res.,* 27:136 (1982).
31. *The Use of Time: Daily Activities of Urban and Suburban Populations in Twelve Countries,* Szalai, A., (Ed.), Mouton, The Hague (1972).
32. Hoogendyk, C.G., Dietz, R.N., D'Ottavio, T.W., and Goodrich, R.W., "Two-zone modeling for air exchange rates and NO_2 source and decay in a southern California residential indoor air quality study," in *Indoor Air '87 Vol. 1,* B. Seifert, H. Esdorn, M. Fischer, H. Ruden, and J. Wegner, (Eds.), (Berlin: Institute for Water, Soil and Air Hygiene, 1987), pp. 333–337.
33. Pellizzari, E.D., Michael, L.C., Perritt, R., Smith, D.J., Hartwell, T.D., and Sebestik, J., "Comparison of indoor and outdoor toxic air pollutant levels in several Southern California communities," Final Report, EPA Contract # 68-02-4544 (Jan. 1989).
34. Alevantis, L.E. and Girman, J.R., "The effects of window-opening on residential ventilation," in *IAQ'89,* (Atlanta: American Society of Heating, Refrigerating and Air Conditioning Engineers, 1989), pp.

35. "Ventilation for Acceptable Indoor Air Quality," ASHRAE Standard 62-1981R, (Atlanta: American Society of Heating, Refrigerating, and Air Conditioning Engineers, Inc., 1989).
36. "Ventilation for Acceptable Indoor Air Quality," ASHRAE Standard 62-1981R, (Atlanta: American Society of Heating, Refrigerating, and Air Conditioning Engineers, Inc., 1989).
37. Persily, A., "Ventilation rates in office buildings," in *IAQ'89,* (Atlanta: American Society of Heating, Refrigerating and Air Conditioning Engineers, 1989), pp.
38. Sheldon, L.S., Handy, R.W., Hartwell, T.D., Whitmore, R.W., Zelon, H.S., and Pellizzari, E.D., "Indoor air in quality public buildings," Volume I and Volume II, U. S. Environmental Protection Agency, EPA/600/6-88/009A and EPA/600/6-88/009B (Aug. 1988).
39. Engelhard Special Chemicals Division. Ozone Converter. Specification SIC-3. Service Information Letter SIL-3. Union, (NJ: Engelhard, 1984).
40. Wallace, L.A., "The Total Exposure Assessment Methodology (TEAM) Study: Summary and Analysis" Volume 1, U. S. Environmental Protection Agency, EPA/600/6-87/002a (June 1987).

CHAPTER 9

The Effect of "Building Bake-Out" Conditions on Volatile Organic Compound Emissions

Charlene W. Bayer

ABSTRACT

The effects of elevated temperatures on volatile organic compound emissions from particleboard and modular office partitions were investigated. This preliminary study employed large-scale environmental chamber techniques in order to achieve precise control over environmental conditions and background volatile organic compound emissions.

INTRODUCTION

Building materials and furnishings are significant sources of airborne volatile organic compounds (VOCs) indoors.[1] Frequently, the levels of VOCs emanating from a product will lessen with increasing product age, as can be seen in Figure 1. Figure 1 shows the VOC emission pattern from (1) a new modular office partition, and (2) the same partition after aging for 4 months at room temperature. The total VOC emissions were reduced 79% after aging.

In an effort to accelerate the natural aging process of building materials and furnishings, "building bake-out" has been suggested. This process involves the elevation of the ambient building temperature for several days and also possibly increasing the building ventilation rate. Girman[2] has reported the results of studies he has conducted where he has "baked-out" buildings. He observed a significant reduction in total VOC concentration after a bake-out. He did find that the individual concentration of selected VOCs showed less of a decrease.

The procedures used to perform the bake-out may significantly influence the effectiveness of the technique to reduce the VOC concentrations. The bake-out is

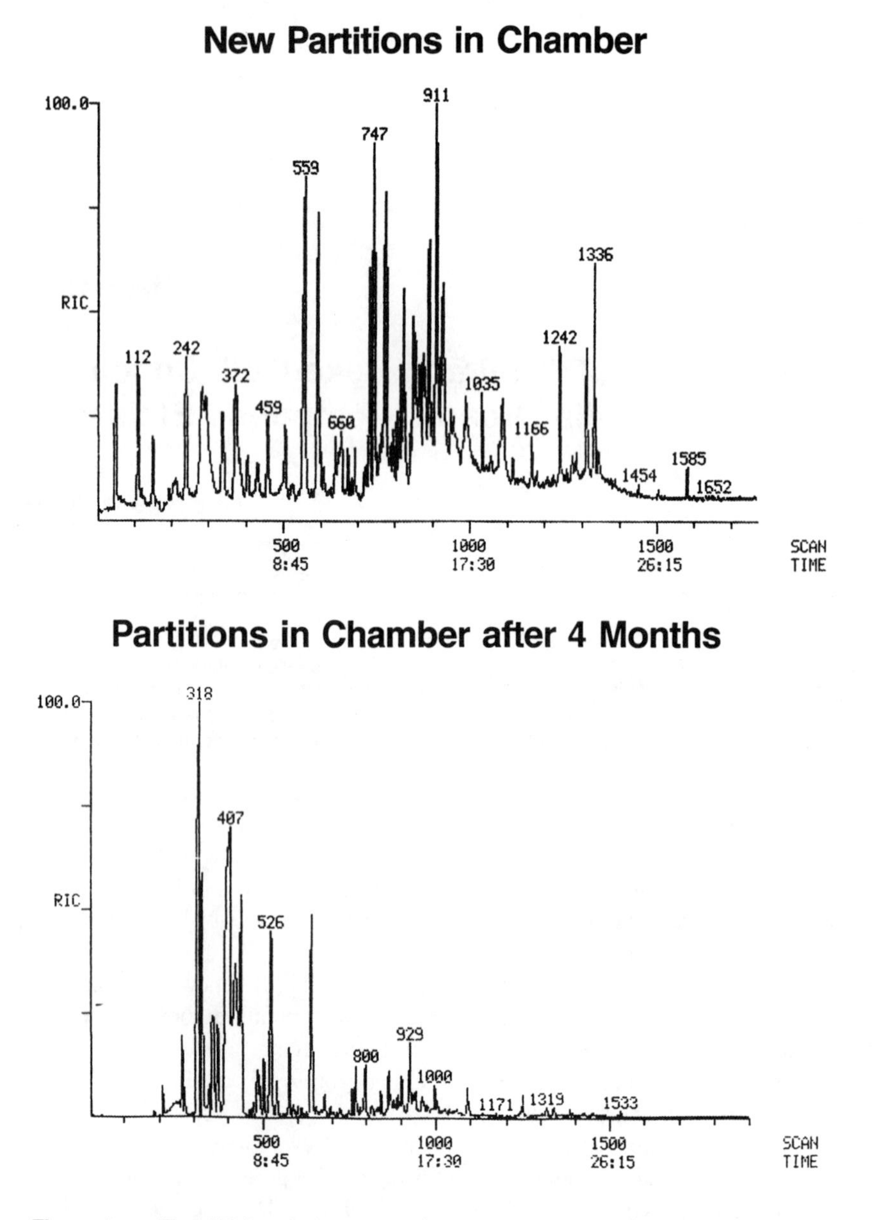

Figure 1. (Top) VOC emission pattern from a new modular office partition. (Bottom) VOC emission pattern of same partition after four months at room temperature.

Table 1. Environmental Chamber Conditions

Volume	28.4 m^3
Temperature	100–120°F (bake-out)
	75°F (pre-bake-out)
	75°F (post-bake-out)
Humidity	50% RH
Air change rate	0.5 ACH
Product loading factor	0.56 m^2/m^3 (partitions)
	1.20 m^2/m^3 (particleboard)

controlled by the duration of the bake-out, the indoor air temperature actually achieved during the bake-out, and the ventilation rate during and after the bake-out. The elevated temperatures must be maintained long enough for the various materials in the building to achieve the elevated temperatures. Girman[2] found that it took approximately 3 d for the internal temperature of the building to reach the bake-out temperature. Sufficient ventilation is required to dilute and remove the airborne pollutants emitted into the atmosphere during the bake-out procedure; otherwise the outgassed compounds will be readsorbed by various building materials and furnishings. If the ventilation rate is too great, then an elevated indoor temperature will not be achieved.

The effect of these and other factors on the VOC emissions from particleboard and modular office partitions, two common building materials and furnishings, were investigated using large-scale environmental chamber techniques. The studied materials were subjected to elevated temperatures while monitoring the VOC emissions. VOC emissions were measured before, during, and after the bake-out procedure.

EXPERIMENTAL

Bake-Out Conditions

The materials to be sampled were loaded into an environmental chamber, 28.4 m^3 in volume, with temperature, humidity, and air change rate control.[3] The environmental chamber conditions employed are listed in Table 1. The pre-bake-out conditions were maintained for 2 to 4 d. Depending on the experiment, the chamber temperature was maintained at 100°F (88°C) and 120°F (99°C) for 3 to 5 d. The post-bake-out conditions were maintained for 2 d. Materials to be tested were loaded into the chamber and conditioned for 24 h prior to emission testing.

Solid Sorbent Tubes

Volatile organic compounds were collected on two different multilayer solid sorbent tubes. Both tubes were prepared in Pyrex glass tubing 17.8 cm in length, 2 mm internal diameter. The first solid sorbent tube used was a three layer tube containing glass beads, Amersorb XE 340, and Tenax®. The preparation of these tubes has been described elsewhere.[4] The second multilayer solid sorbent tube was a four-layer tube containing 0.25 g glass beads, 0.15 g Carbotrap C, 0.20 g Carbotrap, and 0.20 g Carbosieve S-III. The layers separated by clean, unsilanized glass wool plugs. The tubes were baked at 300°C for 19 h while being purged with a stream of 20 ml/min nitrogen. The tubes were cooled to room temperature while being continuously flushed with nitrogen. The tubes were then stored in clean, baked glass containers with PTFE-lined screw caps. These containers were used for transportation of the sampling tubes to and from the sampling site.

Sample Collection

The VOC analytes were concentrated on the solid sorbent sampling tubes by drawing chamber air through the sorbent tube at 200 cc/min for 60 min using personal sampling pumps.

Analysis

The solid sorbent tubes were analyzed by thermal desorption/gas chromatographic/mass spectrometric (TD/GC/MS). This was accomplished with a Tekmar Model 5010 thermal desorber interfaced with a Finnigan OWA 30B GC/MS. The desorber is equipped with a cryofocusing unit to trap volatiles on a length of deactivated aluminum-clad, fused silica capillary tubing (3 m × 0.20 mm I.D.). The fused-silica trap was connected to a DB 624 (J & W Scientific, Inc.) capillary column using a glass butt-end connector. The following conditions were used for thermal desorption: prepurging of the tube for 5 min at 42°C; desorption at 290°C for 10 min; cryotraps cooled to –150°C with liquid nitrogen during desorption and transfer through the transfer line; transfer temperature 290°C for 3.5 min; valve temperature 290°C; injection temperature 290°C; and injection time 0.75 min. The GC separation was begun immediately following flash evaporation from the fused silica trap onto the head of the capillary GC column using the following temperature program: 30°C isothermal for 1 min; 8°C/min to 220°C; isothermal for 10 min. GC/MS transfer oven temperature was 275°C. Mass spectral data were collected between 42 and 300 amu in 0.7 sec. External standards were prepared by injecting known concentrations of the analytes of interest loaded onto solid sorbent tubes as gaseous standards using static dilution bottle techniques.[4] The external standards were analyzed under identical conditions as those used for the sample tubes.

RESULTS AND DISCUSSION

The initial bake-out study was conducted on modular office partitions that had been removed from an office building known to have volatile organic compound contamination. The partitions had adsorbed VOCs from the building in sufficient quantity to have a strong VOC odor similar to that in the building. These partitions were subjected to bake-out conditions in order to determine whether mitigation of the partitions by using bake-out procedures was possible. In this first experiment, the two sampling locations were outside of the environmental chamber. The pre-bake-out period was continued for 4 d to ascertain that constant VOC emissions patterns were occurring. Emissions were highest the first day of the sampling in the pre-bake-out phase, but a constant emission pattern was obtained after the second day. The VOC emission patterns obtained in the 3 d of the pre-bake-out phase are depicted in Figure 2. The temperature inside of the chamber was then raised to 120°F while maintaining the ventilation rate at 0.5 ACH. The emissions of the VOCs increased significantly for the first 2 d of the bake-out. The reconstructed ion chromatograms (RICs) of this phase are shown in Figure 3. On the third day of bake-out, the VOC emissions appeared to lessen, particularly of the higher molecular weight and more polar compounds. This reduction in emissions continued until the sixth day of bake-out when no VOCs were detected, Figure 4. Selected VOCs identified at various sampling intervals during the study are given in Table 2. It was discovered that water was condensing in the sampling lines so that water was being preferentially collected by the sorbent tubes. The water condensation was occurring since the sampling lines were at a cooler temperature than the temperature inside the chamber. An attempt was made to purge the water from the sampling lines with helium prior to collection to alleviate the problem. This did not solve the problem. In this experiment, the VOCs were being collected on the three-layer adsorbent tubes. The adsorbents Tenax® and Ambersorb XE 340 have the disadvantage of being extremely sensitive to high humidity conditions. The Ambersorb beads were swelling to several times their normal size from the collection of the water. The collection media was changed to the four-layer solid sorbent tubes. The adsorbents used in these tubes are much less sensitive to high humidity conditions. These adsorbents also have the advantage of fewer artifacts and easier removal of background contamination prior to sample collection. The use of the four-layer tubes did not alleviate the losses of the more polar compounds. These compounds were probably preferentially being dissolved in the water and were not being adsorbed by the sampling media. It was also hypothesized that VOCs were being lost by attachment to mineral salts plating out on the walls of the environmental chamber.

Modifications in the environmental chamber were made to reduce the losses of VOCs to condensation and plating of the mineral salts. A demineralizer was installed on the water supply to the chamber humidifier. The sampling lines were moved inside the chamber so that the sampling lines and sorbent tubes were the same temperature as the chamber air. The solid sorbent tubes were not characterized for their collection efficiency at the bake-out temperatures for this preliminary study. These modifications appeared to eliminate the VOC losses due to condensation and mineral salt plating.

Particleboard samples were then loaded into the chamber in order to study the effects of elevated temperatures on VOC emissions. The pre-bake-out phases was continued for 2 d, then the temperature was elevated to 100°F for 5 days, and post-bake-out phase was conducted for 2 d. Figures 5 and 6 depict the RICs of the three phases. Selected VOCs detected during the three phases are given in Table 3.

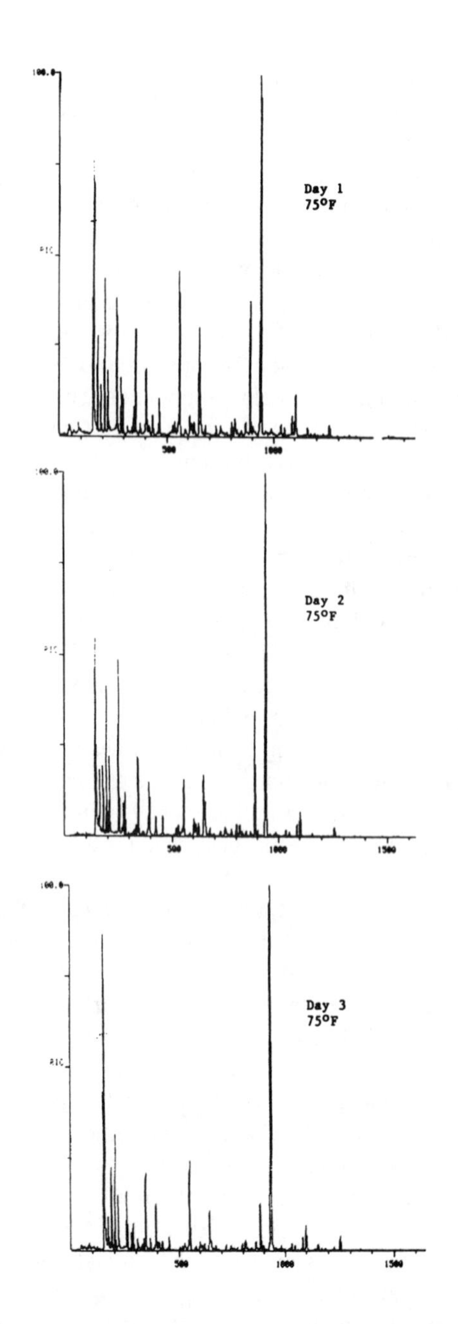

Figure 2. VOC emission patterns from partition before bake-out.

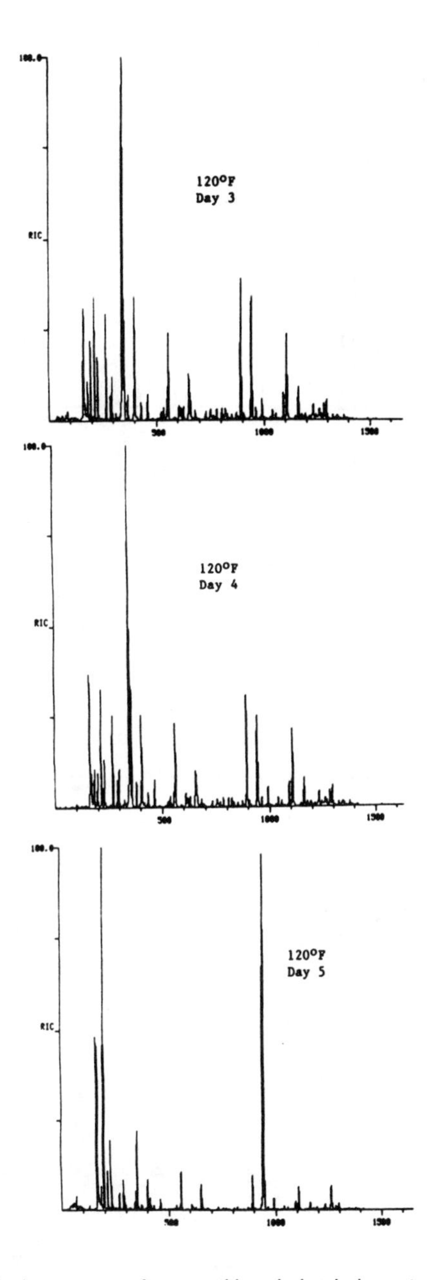

Figure 3. VOC emission patterns from partition during bake-out.

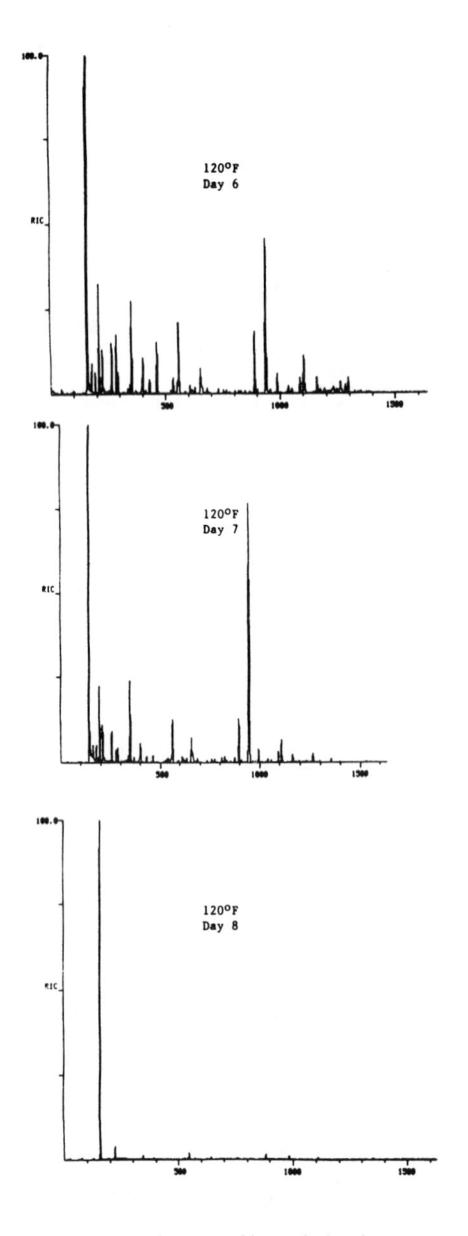

Figure 4. VOC emission patterns from partitions during latter stages of bake-out.

Several higher molecular weight compounds were detected only in the post-bake-out period. Although reductions of certain VOCs did occur during and following the elevated temperatures, certain VOC concentrations such as toluene and hex-

Table 2. Selected VOC Emission Detected From Modular Office Partitions

	Concentration (mg/m³)			
Compound	**Pre1**	**Pre2**	**During1**	**During2**
Ethanol	0.50	0.71	0.78	0.15
1,2-Dichloroethene	1.7	1.4	6.1	BD
Acetone	21	19	0.29	BD
Hexanes	8.3	27	—	1.5
Trichloroethene	0.60	1.1	3.8	BD[a]
1,1,1-Trichloroethane	0.90	1.6	BD	BD
Benzene	7.0	11	19	4.2
Toluene	8.6	22	45	1.5
Methylene chloride	3.2	9.8	BD	7.5
Xylenes	2.8	9.4	9.1	BD

[a] BD is below detection limits.

anes rose during the post-bake-out phases. Polar compounds, such as phenols, were detected only in the pre-bake-out phases indicating that losses of polar compounds were still occurring. Table 4 gives the total VOC concentrations detected during the three phases of the bake-out procedure in millions of counts. This indicated that the bake-out procedure was unsuccessful at reducing the VOC emissions from the particleboard. Total VOC counts were reduced during the period of the elevated temperatures. This could be the result of changes in the adsorption characteristics of the solid sorbents from the elevated temperatures.

The chamber was loaded with another set of modular office partitions removed from a different area of the contaminated building. The VOC emissions from these partitions rose and dropped off during the elevated temperature phase of the study, but total VOC emission levels reported as millions of counts (Table 5) did not show a reduction.

CONCLUSIONS

These preliminary studies of VOC emissions from building materials and furnishings seemed to indicate that elevating ambient temperatures around the products to accelerate the aging process was not successful. In the case of particleboard, it appeared that the elevated temperatures may have allowed the release of higher molecular weight compounds into the atmosphere that were not released at the lower temperatures.

These are very preliminary results. Studies on the effect of bake-out procedures on building materials and furnishings are continuing. The effect on the collection efficiencies of the adsorbents from the elevated temperatures and humidities is continuing. The aging processes of various building materials is being further investigated.

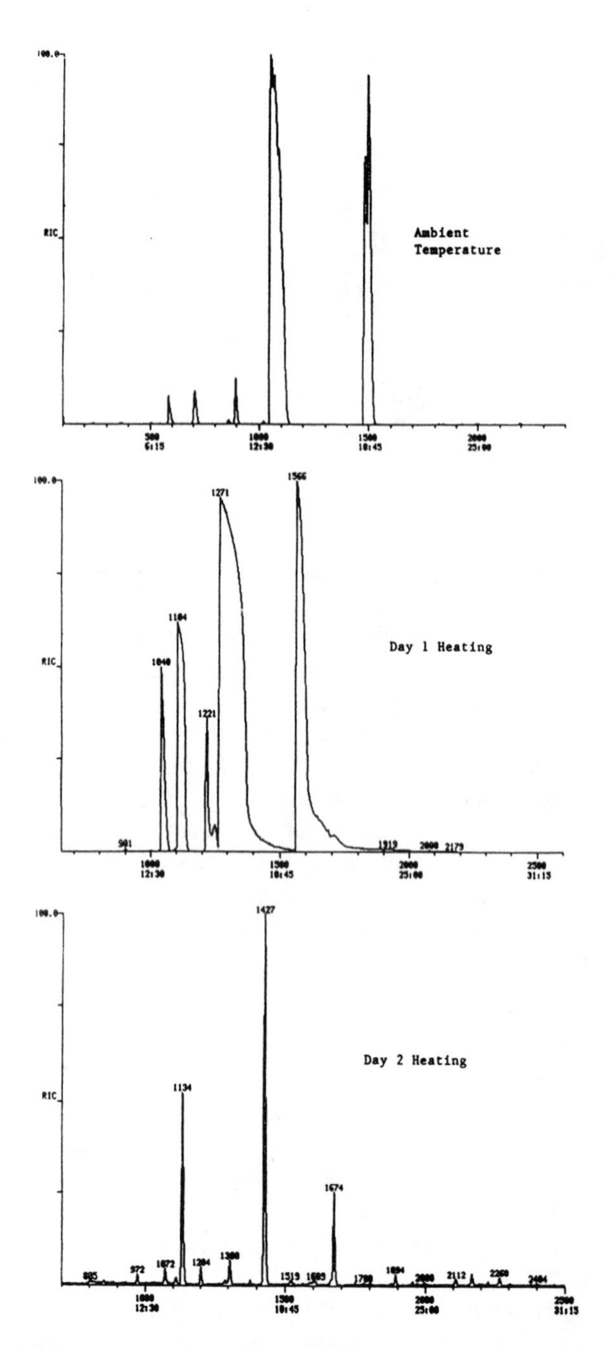

Figure 5. VOC emission patterns from particleboard before and during bake-out.

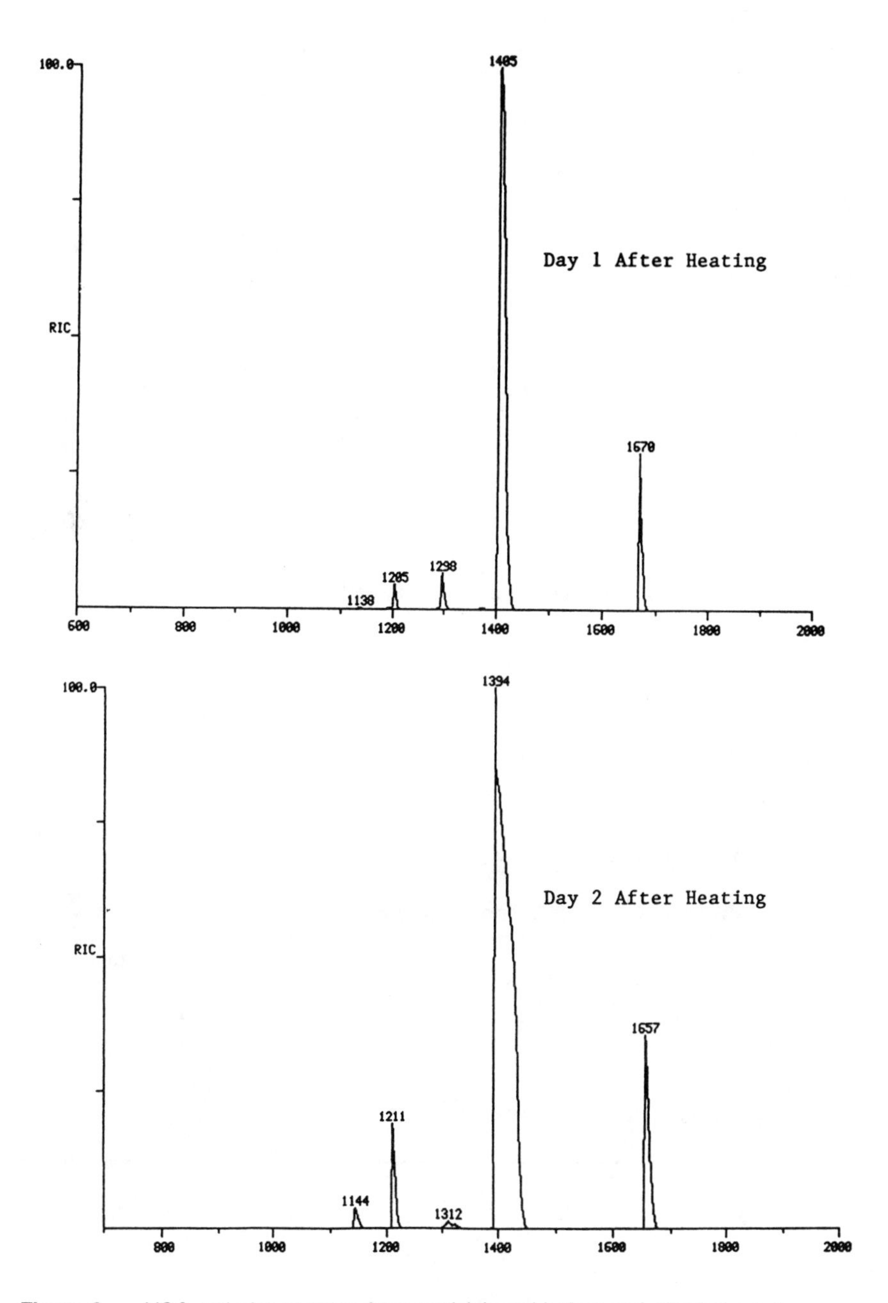

Figure 6. VOC emission patterns from particleboard before and after bake-out.

Table 3. Selected VOC Emissions From Particleboard

	Concentration (mg/m³)					
Compound	**Pre**	**Pre**	**During1**	**During3**	**Post1**	**Post3**
Methylene chloride	15.8	83.4	69.7	42.5	19.9	0.742
Hexanes	29.1	26.8	47.8	9.12	283	501
Ethyl acetate	1.97	52.4	12.8	3.16	25.0	24.5
Benzene	322	48.1	252	46.4	1230	24.5
Toluene	934	31.6	73.5	60.2	340	722
Acetone	0.625	2.61	1.18	2.62	—	0.472
Benzaldehyde	9.86	7.97	0.750	BD	BD	BD
Ethylbenzene	0.520	4.05	19.2	BD	BD	BD
Trichloroethene	BD	BD	BD	BD	BD	1.26
3,4-Dichloro-2,5-furanedione	BD	BD	BD	BD	BD	0.244
Phenol	BD	4.18	BD	BD	BD	BD
4-Methylphenol	BD	0.996	BD	BD	BD	BD

Table 4. Total VOC Concentrations From Particleboard

	Millions of counts
Pre-bake-out day 1	2.6
Pre-bake-out day 2	4.0
Bake-out day 1	0.04
Bake-out day 2	0.04
Bake-out day 3	0.06
Post-bake-out day 1	1.6
Post-bake-out day 2	3.3

Table 5. Total VOC Concentrations From Partitions

	Millions of counts
Pre-bake-out	0.23
Bake-out day 1	0.73
Bake-out day 2	0.40
Bake-out day 3	0.31
Bake-out day 4	0.30

REFERENCES

1. Molhave, L., "Indoor Air Pollution Due to Organic Gases and Vapors of Solvents in Building Materials," *Environ. Int.*, 8:117–127 (1982).
2. Girman, J. et al., "Bake-out of a New Office Building to Reduce Volatile Organic Concentrations, Paper 89-80.8." Presented at Anaheim, CA: the 82nd Annual Meeting of the Air and Waste Management Association (June 1989).
3. Black, M.S. and Bayer, C.W., Formaldehyde and VOC exposures from some consumer products, in *Proceedings of IAQ'86, Managing Indoor Air for Health and Energy Conservation,* (Atlanta: ASHRAE, 1986), p. 454.
4. Winberry, W.T. et al., "Method IP-1B, Solid Sorbent Tubes," in *Compendium of Methods for the Determination of Air Pollutants in Indoor Air,* Draft (September 1989).
5. Bayer, C.W., Black, M.S., and Galloway, L.M., "Sampling and Techniques for Trace Volatile Organic Emissions from Consumer Products," *J. Chromatogr. Sci.*, 26:168–173 (1988).

CHAPTER 10

Air Cleaners for Indoor Air Pollution Control

A.S. Viner, K. Ramanathan, J.T. Hanley, D.D. Smith, D.S. Ensor, and L.E. Sparks

ABSTRACT

Indoor air pollutants include both particles and gases, and different technologies are required to control these pollutants. An experimental study was conducted to evaluate the performance characteristics of currently available control technologies. One aspect of the study was to evaluate the particle-size dependent collection efficiency of seven commercially available devices for control of particles: one common furnace filter, four industrial-grade filters, and two electronic air cleaners (EACs). The furnace filter had negligible effect on particles in the size range 0.1 to 1 mm (i.e., those that penetrate deep into the human lung). The industrial-grade filters, which had ASHRAE ratings of 95, 85, 65, and 40%, exhibited a minimum efficiency at approximately 0.1 mm, which was substantially lower than the ASHRAE efficiency. Of the two EACs, one was essentially a furnace filter with a high-voltage electrode while the other was similar to an industrial electrostatic precipitator (ESP). The furnace-filter type of EAC reached a maximum efficiency of 30% at low flowrates (7 m^3/min); however, it had Ra negligible effect at higher flowrates (14 and 20 m^3/min). The ESP-like EAC exhibited efficiencies from 80 to 90% over the entire size range at low to moderate flow rates. At the highest flowrate, a minimum efficiency of 60% was detected at 0.35 mm. Measured ozone emission rates for the EACs were used to estimate a worst-case ozone exposure in a typical residential environment. The worst-case scenario yielded a maximum concentration of 60 ppb. Actual concentrations would be much lower.

Another aspect of the study was to evaluate the suitability of commercially available carbon-based sorbents for removing low concentrations of volatile organic compounds (VOCs). A laboratory experiment was conducted to measure the capacity of three different carbons (wood-, coal-, and coconut-shell-based carbons) for three different VOCs (benzene, acetaldehyde, and 1,1,1-trichloroethane) at low concentrations (100–200 ppb). Measured capacities ranged from 10^{-8} to 10^{-6} g-mol/g carbon. Model calculations based on a challenge concentration of 150 ppb and a breakthrough concentration of 50 ppb indicated that commercially available 15 cm (6 in) thick in-duct carbon filters would have a bed life on the order of minutes.

INTRODUCTION

The growing awareness of indoor air pollution has been accompanied by a growing interest in air cleaners to remove pollutants from the air. Although air cleaners have been on the market for many years, there is very little information available on how well these devices work. Within the past few years, the U.S. Environmental Protection Agency (EPA) has begun research programs to evaluate various aspects of indoor air pollution. One such program is being conducted at the Research Triangle Institute (RTI) to evaluate indoor air cleaner technologies.

Indoor air pollution can be divided into two categories: gases and particles. Gaseous pollutants include VOCs, combustion products, and other substances. The VOCs in indoor air are the result of emissions from clothing (e.g., chemicals used in dry cleaning), building materials (e.g., plywood), carpeting, and upholstery. Products of combustion result from cooking and gas heat and include carbon dioxide and carbon monoxide, among other gases. Any or all of these gases may be present in indoor air at one time. This study focused on removal of VOCs from indoor air. The novel aspect of this problem in the context of indoor air is the low concentrations that are encountered. Typical concentrations that may be found in an indoor environment are on the order of 100 ppb. Because no isotherm data have been published for sorbents at these concentrations, it is not clear how well air cleaners will work in an indoor environment.

Airborne particles are the result of cooking, smoking, infiltration of dust and pollen from outdoors, and everyday activities in the home or office. While all particles can be a nuisance, the particles of most concern are those with a diameter between 0.1 and 1 μm. These minute particles penetrate deep into the human lung and may be of concern from a health standpoint.

The purpose of this report is to describe the results of our research program to evaluate five air filters and two electronic air cleaners (EACs), and to evaluate three different carbon sorbents that are typical of those used for removal of vapors from indoor air. The experimental approach and the results obtained are described below.

PARTICLE AIR CLEANERS

There are two main types of indoor air cleaners: in-duct and room cleaners. In-duct cleaners are installed within heating, ventilating, and air-conditioning (HVAC) systems. Room air cleaners contain a filter and perhaps an absorber with a self-contained fan. The control approach used in both cases includes fibrous filtration, electrostatic precipitation (ESP), and/or electrostatically augmented filtration. Vapors are usually absorbed by activated carbon. A more complete description of the hardware may be found in the *ASHRAE Handbook*.[1] Recently, the Association of Home Appliance Manufacturers published a standard test procedure[2] for evaluating room air cleaners. However, only in-duct air cleaners were considered in this research.

Filter Test Results

The filter efficiency tests were conducted in the apparatus illustrated in Figure 1. The test section of the system has back-to-back stainless steel filter holders that can accommodate filters as large as 60 × 60 cm (2 × 2 ft). System ducting is 20 cm (8 in) diameter polyvinyl chloride (PVC) both upstream and downstream of the test section. Thus, for a given flow rate, the air velocity through the ductwork is much greater than that through the test section. The resulting turbulence enhances mixing of the particles in the gas stream and improves accuracy in velocity measurements and particle sampling.

The test section is maintained at 1 cm H_2O positive pressure to prevent inleakage of aerosol, which would bias the results. In addition, the test section is enclosed within a Class 100 clean room. Depending upon the test and the allowable background aerosol concentration, the test air may be drawn either from the clean room airflow or from the ambient laboratory air. Aerosol penetration ratios as small as 10^{-8} have been measured with this apparatus.

The aerosol sample taps are located just upstream of the test section and also sufficiently far downstream of the test filter to allow for the complete mixing of any penetrating aerosol with the entire airstream. Running the downstream ducting back under the filter holders locates the challenge and penetrating aerosol taps near each other, thereby allowing short sample lines and reducing particle losses. The losses in the ductwork were found to be insignificant for the particle sizes of interest.

Aerosol instruments used for the tests included a PMS, Inc. Laser Aerosol Spectrometer (LAS-X), a Climet 226/8040 Optical Particle Counter (OPC), and a TSI Differential Mobility Particle Sizer (DMPS). The laser counter was used to measure the particles from 0.009 to 3 μm, the OPC from 0.3 to 3 μm, and the DMPS from 0.011 to 0.457 μm.

Airflow was monitored with a digital thermal anemometer probe placed at the center of the duct. Duct traverses and pitot probe measurements were used to determine the flowrate. The pressure drop of the air cleaner was measured with an included manometer.

An aerosol generated by nebulizing aqueous solutions of potassium chloride (KCl) was used as the particle challenge. The mean diameter of the dry residue of KCl was controlled by the strength of the solution and was varied during the test to cover the range from 0.01 to 3 μm.

Five filters were tested at three flowrates: 7, 14, and 28 m^3/min. One filter was a common household furnace filter and the other four were industrial grade filters with ASHRAE ratings of 40, 65, 85, and 95% efficiency.

An example of the results obtained for a furnace filter is shown in Figure 2. Even at the lowest flowrate (7 m^3/min), the filter had a negligible efficiency for

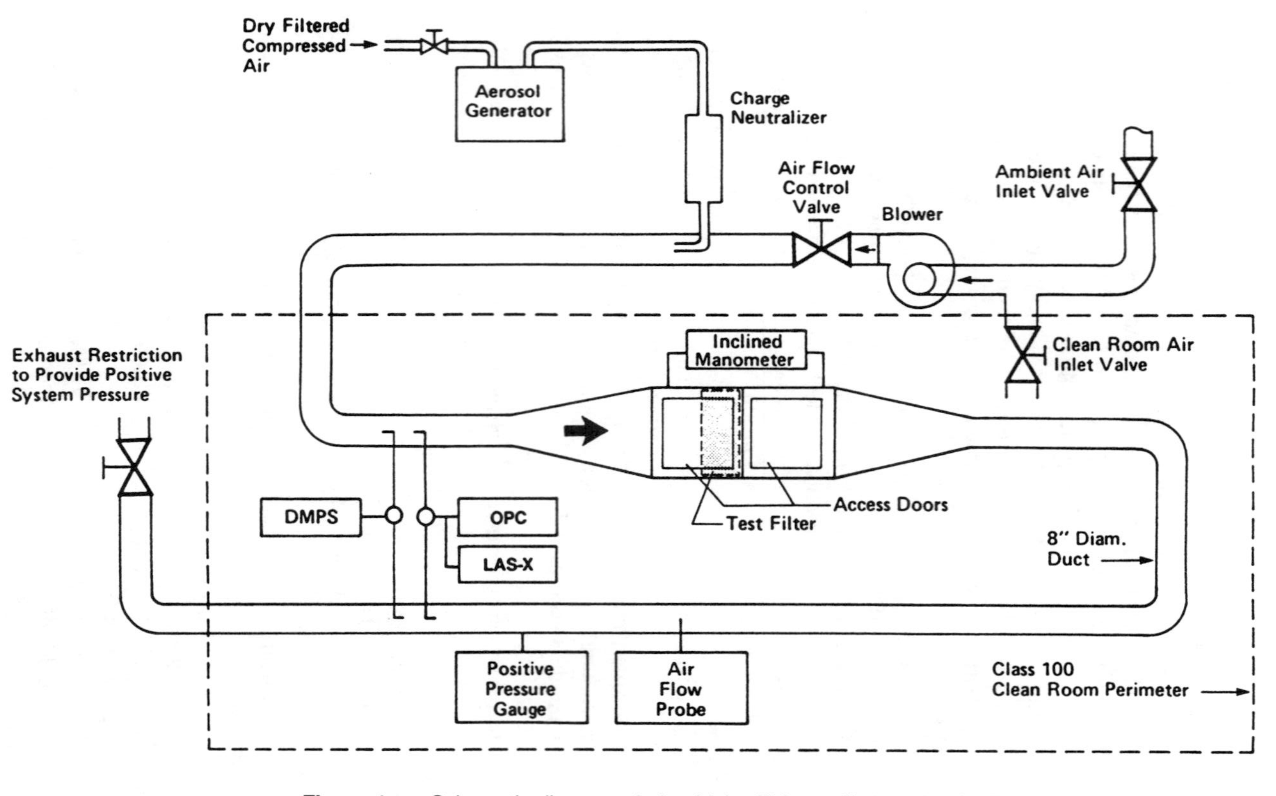

Figure 1. Schematic diagram of ultra-high efficiency filter testing facility.

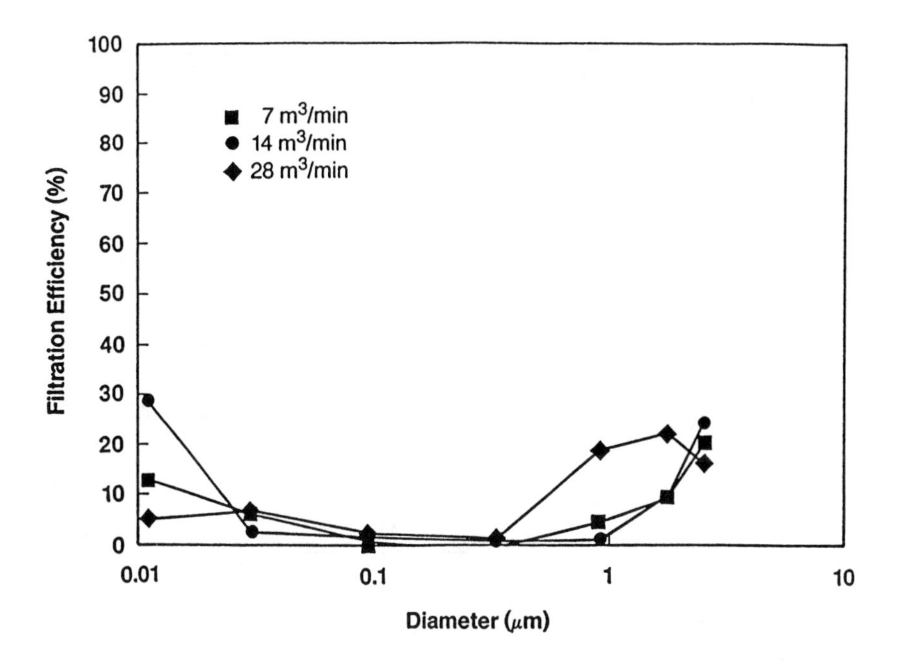

Figure 2. Filtration efficiency vs. particle diameter for the furnace filter.

removal of small particles (i.e., those smaller than 1 μm). The particles between 0.1 and 1.0 μm deposit deep within the human lung. At higher flowrates (14 and 28 m^3/min) the collection efficiency decreased. These results reflect the fact that furnace filters are designed to prevent dust balls from clogging the heat transfer surfaces of furnaces and air conditioners and are not intended to protect human health.

Particles between 0.1 and 1.0 μm also affect equipment, especially electronic equipment.[3] Particles in this size range deposit on vertical as well as horizontal surfaces. A large fraction of submicron particles consist of water-soluble salts. When the relative humidity rises above the deliquescence point of these salts, they can cause current leakage, shorts, and corrosion. Resulting failures cost U.S. companies millions of dollars per year and are of concern wherever highly integrated electronic equipment operates.

Industrial-grade filters are somewhat more effective at collecting small particles. Figure 3 represents the size-dependent efficiency for four filters at their rated flowrate (14 m^3/min). All of the curves show a sharp dip in the efficiency curve, with a minimum near 0.1 μm. The general shape of these curves is typical of filters. For particles larger than 1 μm, the dominant collection mechanisms are interception and impaction. For particles smaller than 0.1 μm, diffusion is the dominant collection mechanism. Filter efficiencies are always worst for particle sizes between 0.1 and 1.0 μm. An important observation about the data is the

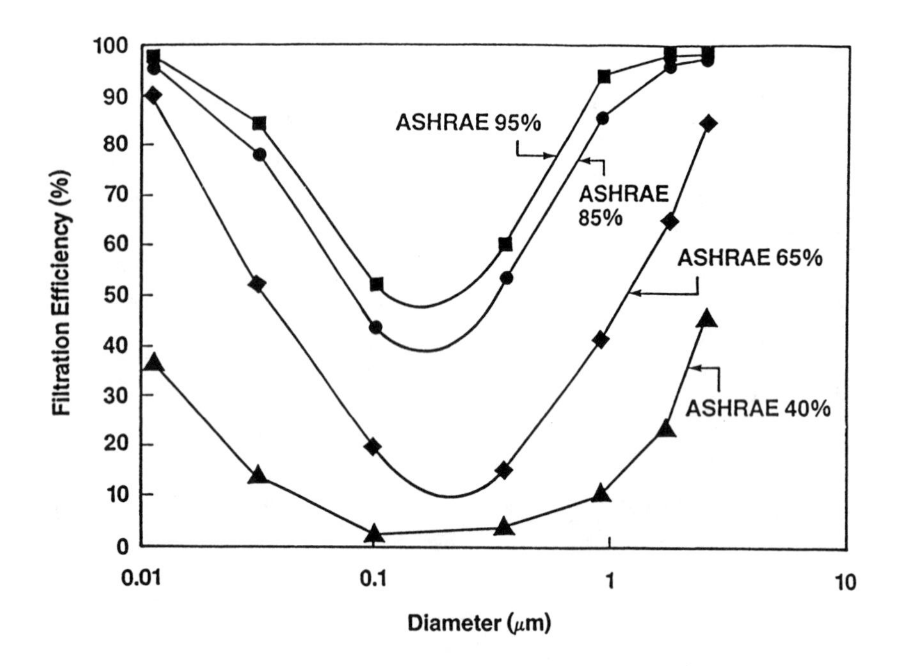

Figure 3. Filtration efficiency vs. particle diameter for the four ASHRAE-rated filters at a nominal flow rate of 14 m^3/min.

discrepancy between the size-dependent efficiencies and the ASHRAE efficiencies. As mentioned above, the ASHRAE test procedure is biased toward large particles. While a filter may collect 95% of particles larger than 1 μm, as many as 50% of particles with a diameter of 0.1 μm penetrate the filter. This points out the weakness of relying on a single number for designation of efficiency. Nevertheless, the industrial filters (with the possible exception of the ASHRAE 40% filter) would serve as good collectors of pollen and other natural aerosols, which tend to have large diameters (5 μm and larger).

EACs

Another way to remove particles from an airstream, either as an adjunct to or in place of filtration, is electrostatics. Several devices have been developed to incorporate electric fields into filters, and some devices combine an ionizing section as well to charge incoming particles. On the other end of the spectrum is the conventional wire-plate ESP in which there is no filtration medium and particles are removed entirely by the charging and electrostatic collection of aerosol particles.

One design of EAC is illustrated in exploded view in Figure 4a. This design, hereafter referred to as the "furnace filter design", consists of high-voltage wires sandwiched between a lightweight, open foam collection medium. Metal grilles on

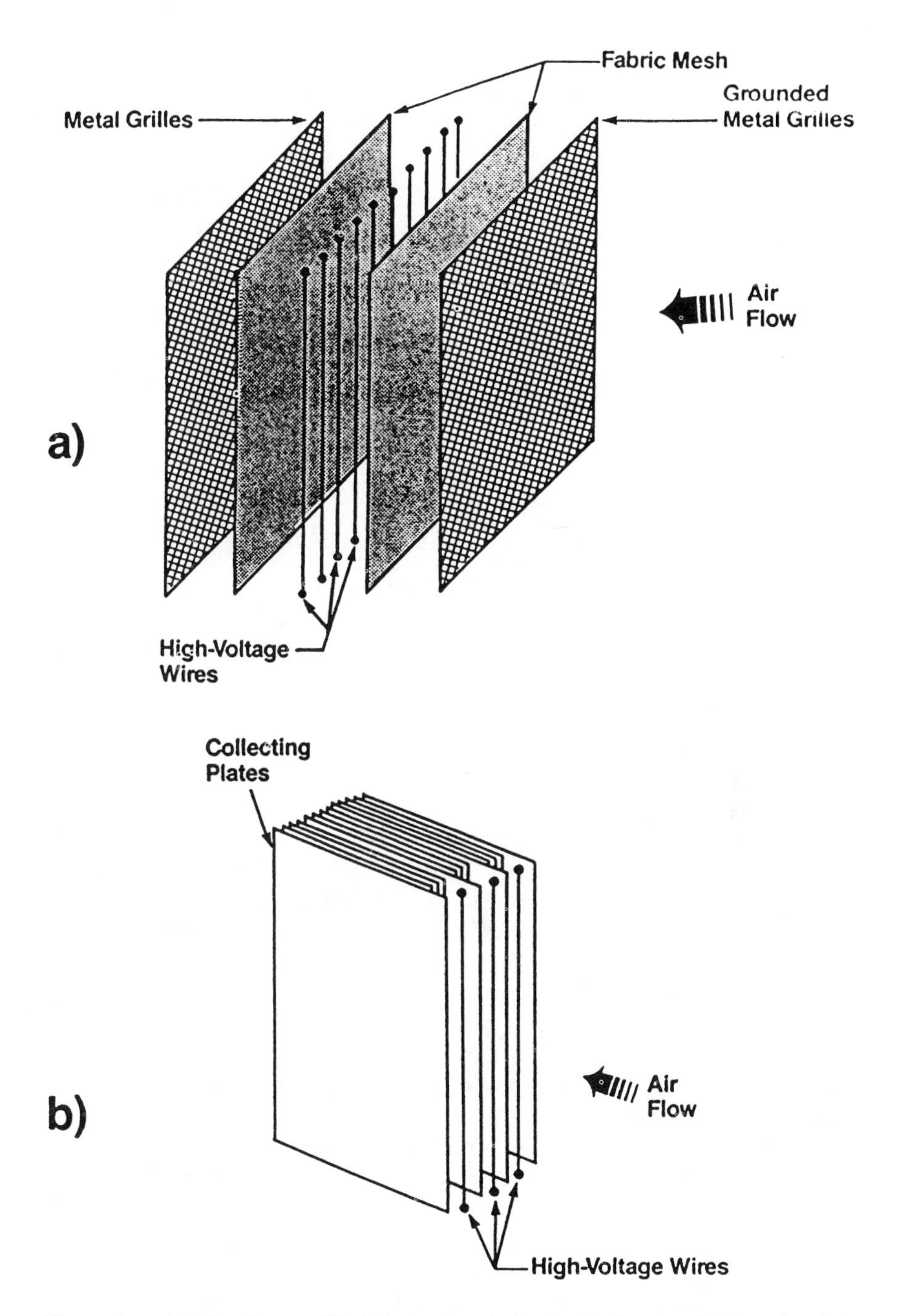

Figure 4. (a) The "furnace filter" design for electronic air cleaners; (b) the wire plate design.

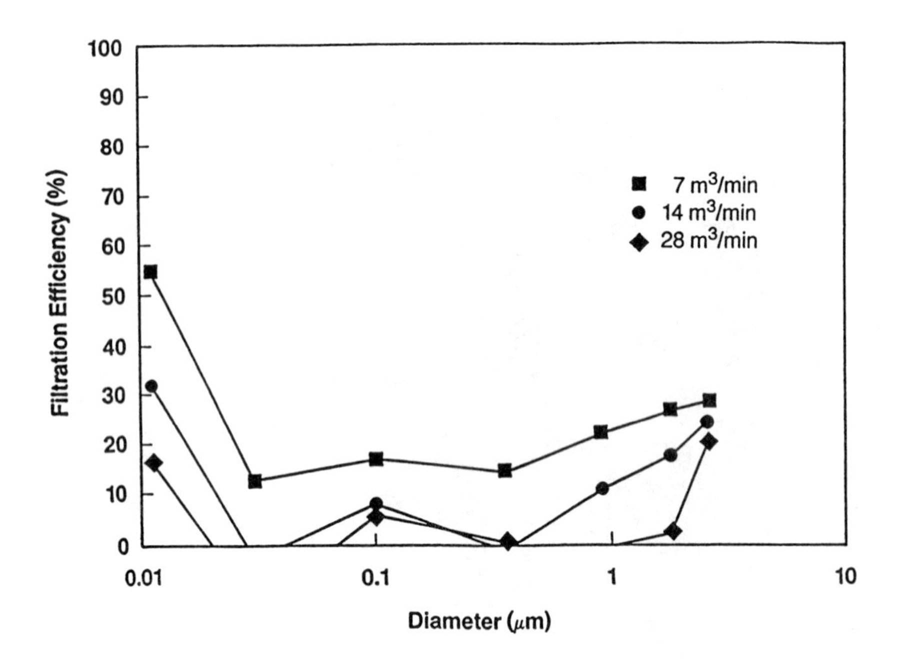

Figure 5. Filtration efficiency vs. particle diameter for Electronic Air Cleaner No. 1.

the outside of the foam serve as ground planes for the electric field. Nonconductive spacers maintain a separation of approximately 12.5 mm between the high-voltage wires and the metal grilles, for a total grille-to-grille width of 25 mm. The unit is designed to replace a conventional furnace filter and thus is easy to install. For safety, the power control unit for the device includes a flow-sensor switch so that the high-voltage current is only on while air is flowing through the device.

The size-dependent collection efficiency of this air cleaner is shown in Figure 5 for three different airflow rates. Except for the smallest particle size at the lowest flowrate, the collection efficiency was consistently below 30%. At moderate to high flowrates (14 and 28 m^3/min) the collection efficiency was even worse. By comparison with the results obtained for a conventional furnace filter (Figure 2), it appears that, for all practical purposes, the use of electrostatics does not improve the efficiency of the filter. As with a conventional furnace filter, this device may be appropriate for removal of large dust balls and thereby provide some protection to the heat-transfer surfaces of a furnace/air conditioner; however, it does not appear to be useful in removing particles that are most likely to affect human health.

A more rigorous design for an electronic air cleaner is shown in Figure 4b. This wire-plate design is very similar to that of ESPs used in industrial plants to control dust emissions. It consists of a parallel array of electrically grounded plates spaced approximately 0.5 cm apart. Airborne particles entering the device acquire a

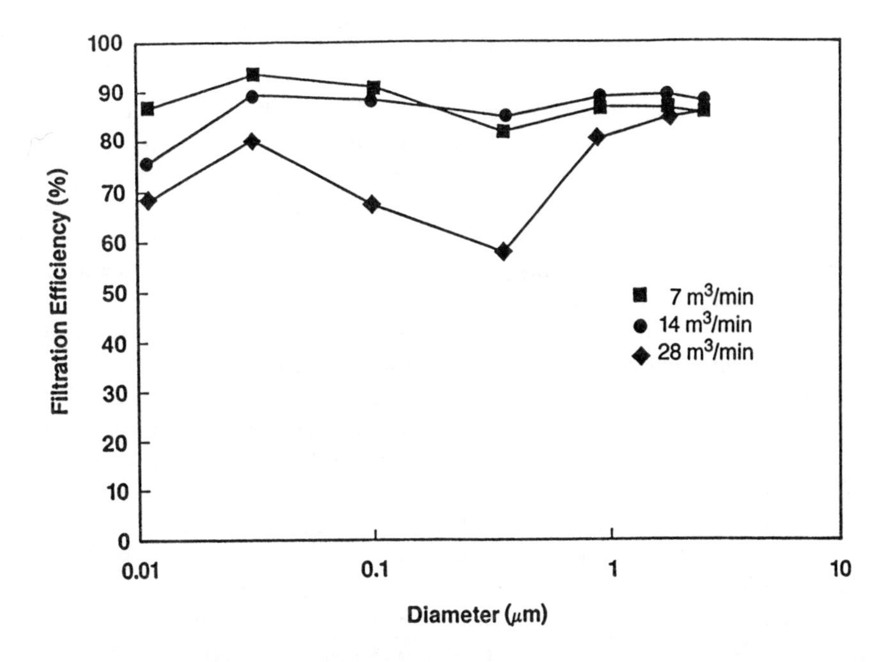

Figure 6. Filtration efficiency vs. particle diameter for Electronic Air Cleaner No. 2.

charge from the ions generated by the high-voltage electrode. The charged particles then travel through the channels bounded by the grounded plates, which are 12 cm long. The length of the channel provides sufficient residence time to attract and collect the charged particles onto the plates.

The filtration efficiency of this device is shown in Figure 6 for three flowrates (7, 14, and 28 m^3/min). At low and moderate flowrates (7 and 14 m^3/min), the efficiency ranged from 75 to over 90% over the entire size range. At the highest flow rate there is a dip in the efficiency curve at 0.35 μm. The decrease in efficiency with increasing flowrate was a result of the reduced residence time in the collection section of the air cleaner. Overall, however, this EAC performs quite well when compared with the industrial filters.

In comparing the performance of EACs and industrial filters, it should be noted that, up to a point, the efficiencies of filters improve as they "load dust". The converse is true for EACs. As their plates load, they become less efficient. Routine inspection and maintenance is even more important with EACs than with filters. (With filters, the pressure drop provides an easily monitored indicator of loading.)

Ozone Emissions from EACs

High electron energies, such as those found in ESPs, induce a number of chemical reactions in gas molecules surrounding electrodes. These gaseous reactions are fundamental steps inherent in electrical breakdowns rather than incidental side effects. Ozone is one of the products that result from electronic excitation of oxygen.

The hazardous nature of ozone has prompted several studies of ozone generation by electrostatic discharges over the last 50 years. Castle et al.[4] conducted one of the more rigorous studies by quantifying ozone generation in a cylindrical ESP using positive corona. Their results confirmed the work of White and Cole,[5] which showed that the ozone production rate is proportional to the power input to the electrode. Based on these results, Castle outlined a set of design criteria for electronic air cleaners to minimize the production of ozone, the most important of which was to use a positive polarity for the discharge (corona) electrode. Virtually all electronic air cleaners designed for indoor air applications use positive corona and are otherwise designed to minimize ozone generation while maximizing collection efficiency. Nevertheless, a small amount of ozone is still generated. The objective of this task was to quantify the ozone generation rate and to estimate the effect on indoor air quality.

The ozone generation rate was measured for both the "furnace filter" EAC and the wire-plate model. The apparatus used to measure ozone emissions is shown schematically in Figure 7. An ultrasonic humidifier was used to inject a water aerosol into room air. The water droplets were evaporated in a mixing chamber, after which the air was introduced into the ozone generation section where the air cleaner was placed. Ozonated air exiting from the air cleaner was funneled into a 10 cm diameter glass pipe through a Teflon®-lined transition section. The glass pipe was 2.4 m long to provide adequate mixing of the ozone in the turbulent air stream. Ozone concentration, air velocity, and dewpoint were measured at the exit of the glass pipe. Given flat profiles for both concentration and air velocity, the total ozone generation rate (in micrograms per second) was calculated from the measured values.

The results obtained from these measurements are summarized in Figure 8, which presents ozone generation rate as a function of humidity at the measured current (i.e., 1.3 mA) for both air cleaners. It appears that the effect of humidity on ozone generation is negligible below 50% relative humidity (R.H.); however, there is a 25% decrease in generation rate as the R.H. increases from 50 to 80%. Interestingly, ozone generation from the furnace-filter EAC was substantially lower than that from the wire-plate EAC. However, given the poor performance of the furnace-filter EAC for removing respirable particles, we shall focus on the wire-plate EAC.

A worst-case estimate of potential ozone exposure can be obtained by considering the use of the wire-plate air cleaner in a typical residential application. A simple dilution model can be used to estimate the steady-state concentration of ozone:

$$C = \frac{q\, C_{outside} + S / V}{q + k} \tag{1}$$

where

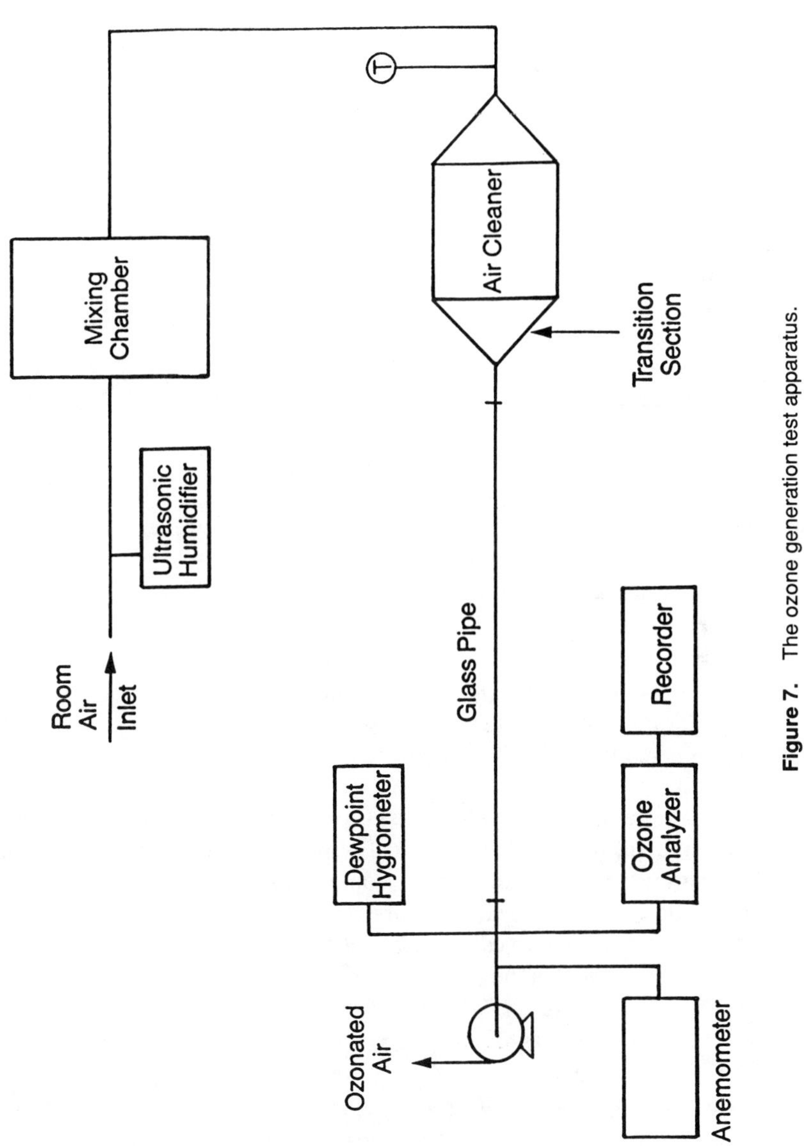

Figure 7. The ozone generation test apparatus.

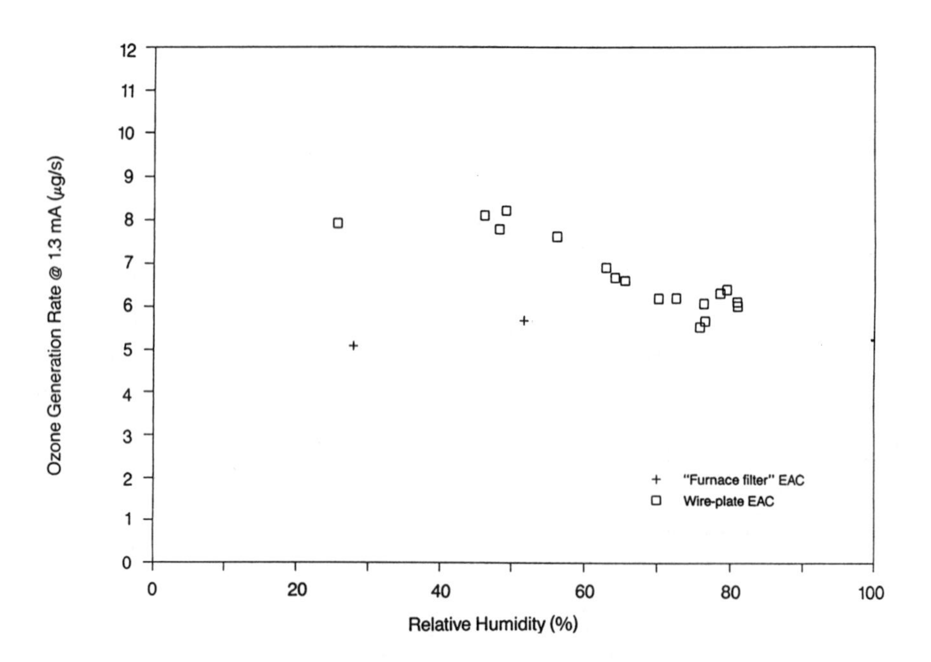

Figure 8. Measured ozone generation rates from commercial air cleaners.

C = ozone concentration (mg/m^3)
q = air exchange rate (h^{-1})
$C_{outside}$ = ozone concentration in the outdoor air (mg/m^3)
S = rate at which the pollutant is generated (mg/h)
V = volume of the house/room (m^3)
k = rate of removal due to processes other than ventilation (h^{-1})

For this worst-case scenario, dilution is the only mechanism for reducing the ozone concentration, so that $k = 0$ (in actual settings, k is nonzero). In addition, we wish to consider only the contribution from the air cleaner, so we will ignore the concentration of ozone in the outside air ($C_{outside} = 0$). Finally, we will consider a fairly "tight" house with a typical volume (456 m^3, equivalent to a 187-m^2 house with 2.44-m ceilings), so that the air exchange rate is fairly low (i.e., $q = 0.4\ h^{-1}$). The ozone generation rate for the wire-plate EAC is approximately 6 mg/s or 21.6 mg/h. Given these parameters, Equation 1 yields a steady-state ozone concentration of 0.118 mg/m^3 (approximately 60 ppb). For comparison, the OSHA Permissible Exposure Limit for an 8-h period is 0.196 mg/m^3. Note that the calculated value (0.118 mg/m^3) is based on continuous ozone emissions; however, air cleaners only operate intermittently. In addition, we have neglected any destruction mechanisms that would further reduce the concentration. Thus, it appears that in actual applications the ozone emissions from electronic air cleaners would have a negligible effect on indoor air quality.

Control of VOCs

There is a wide variety of sources of VOCs in indoor air. These sources vary tremendously depending on the nature of the facility, but often include contributions from outdoor air that enters the building through leaks and through the HVAC system. Other typical sources include combustion, smoking, building materials, and office machines. Various studies of indoor air quality have identified more than 250 organics at levels greater than 1 ppb[6] with total VOC concentration expected in the few hundred parts per billion range and not exceeding a few parts per million.[7,8]

A variety of methods have been proposed for control of VOCs. The most obvious methods are source removal and ventilation. These methods have received the most attention as control strategies. However, source removal is not always practical, and increasing the ventilation rate can result in prohibitively high HVAC costs. Other possible control strategies include adsorption, absorption, incineration, and catalytic conversion. All have demonstrated applicability in conventional pollution control applications and could be incorporated into an HVAC system with minimal retrofit. However, adsorption onto carbon seems to be the most popular technique currently in use for indoor applications.

The majority of adsorption systems currently marketed for indoor air pollution applications deal with odor control. These systems consist primarily of in-duct carbon filters. Quantifiable performance data are not available for any of the systems surveyed. The carbon in each unit is assumed to be spent when occupant complaints increase. Obviously, this method of bed lifetime estimation has drawbacks, particularly for potentially toxic compounds that have relatively high odor thresholds. Adsorption isotherm data in the low concentration regime (<10 ppm) are not publicly available for even simple compounds on any carbons. This led to our effort to obtain adsorption isotherms in the parts per billion range in order to determine the applicability of activated carbon adsorption for indoor air pollution control.

Experimental Design

Three types of activated carbon were tested: a coal-based carbon, a wood-based carbon, and a coconut-shell-based carbon. The activation method for all of these carbons was by steam, and surface areas ranged from 1050 to 1150 m^2/g. The pollutants chosen were relatively simple, low-molecular-weight compounds representative of three different classes of organic compounds: benzene (aromatic hydrocarbon), acetaldehyde (oxygenated hydro-carbon), and 1,1,1-trichloroethane (halogenated hydrocarbon).

The experimental setup is shown in Figure 9. It consisted of two mass flow controllers with flow ranges of 0 to 1000 and 0 to 100 cm^3/min, a microbalance, a thermostated water bath for the permeation tubes, and a balance jacket.

The microbalance used for the adsorption studies is capable of weighing up to 100 g with an accuracy of ±0.5 μg. This device may be operated under low-temperature, high-vacuum conditions for the determination of BET surface area

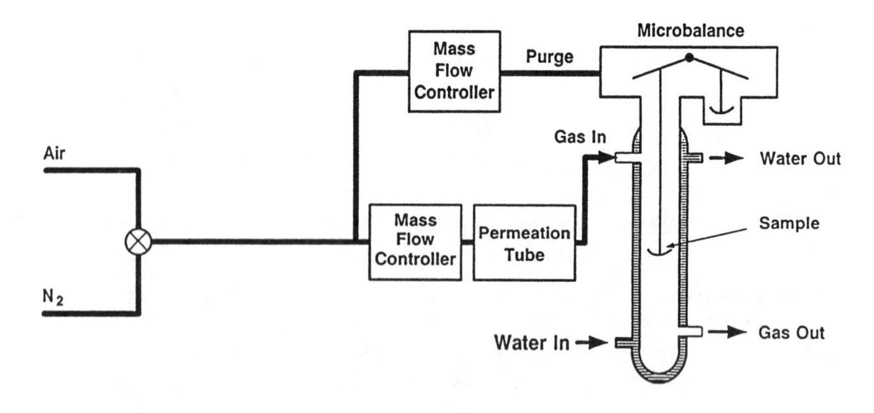

Figure 9. Schematic of the carbon adsorption experimental setup.

and is ideally suited to make adsorption isotherm measurements under different conditions of pressure and temperature.

Isotherms were measured in the dynamic (flow) mode at atmospheric pressure to reduce experimental time and to measure adsorption under conditions representative of real-work situations. The carbon sample (150 to 200 mg) was degassed under flowing nitrogen while being heated with an infrared lamp. Degassing was assumed to be complete when the weight remained constant for 10 to 15 min. The sample was then allowed to cool in nitrogen. Next, the nitrogen was switched off and an airstream was introduced through the permeation tube. The challenge concentrations were generated with clean, dry airstreams flowing over permeation tubes with certified permeation rates. Flow rates were adjusted to give the desired concentrations. All gases had purities greater than 99.999%.

Equilibrium was assumed when the weight gain (on the 100-μg scale) leveled off.

Results

The results for the uptakes of organic pollutants on the various carbons are summarized in Table 1. It is apparent that the capacities (g-mol/g carbon) do not vary much with either compound or carbon in the concentration range studied. Capacities ranged from 10^{-8} to 10^{-6} g-mol/g. Acetaldehyde uptake was about two to three times higher than for the other compounds studied.

While isotherm data are useful for carbon screening, they do not provide direct practical information on the applicability of the technique. For this reason, the weight of carbon required was estimated using a method based on carbon capacity recommended by the manufacturer of two of the carbons[9] and a fixed bed breakthrough model developed at RTI. These methods were used to substantiate the results. The superficial velocity of gas across the filter was assumed to be 12.7 cm/s, corresponding to a flow of 2.8 m^3/min (100 cfm) across a 60 × 60 cm (2 × 2 ft) filter. An exit concentration of 50 pbb was set as the level corresponding to useful bed life.

Table 1. Summary of Adsorption Results

Compound	Concentration (ppb)	Uptake[a] (× 10^7)		
		Wood-base	Coal-base	Coconut shell-base
Acetaldehyde				
mol CH_3CHO/g C	119	3.39	3.52	10.0
	153	4.17	10.00	10.0
Benzene				
mol C_6H_6/g C	101	2.42	1.24	4.30
	119	4.30	6.47	5.60
	176	10.00	7.27	6.33
1,1,1-Trichloroethane				
mol CH_3CCl_3/g C	115	0.749	1.15	2.04
	183	1.33	2.80	1.19

[a] Each number represents an average of 3 to 4 data points.

Table 2. Calculated Carbon Bed Lifetime (15 cm)

Gas	Concentration (ppb)	Lifetime (min)		
		Wood	Coal	Coconut shell
Benzene	100	1.02	0.52	1.82
	120	1.54	2.29	2.01
	176	0.24	1.76	1.54
Acetaldehyde	119	1.22	1.26	7.14
	153	1.16	2.79	2.79
1,1,1-Trichloroethane	115	2.78	0.43	0.76
	183	0.31	0.65	0.45

The first method was applied to the data for all three compounds with challenge concentrations varying from 115 to 183 ppb. The lifetime for 15 cm (6 in.) thick filters (such as those typically marketed for odor control applications) was determined. The results in Table 2 show that breakthrough occurs rather quickly, bringing into question the applicability of a once-through filter even in the absence of other pollutants or humidity effects, both of which would cause performance to deteriorate further.

The second method, the RTI program, was applied only to the benzene data. MA regression analysis was performed to obtain a linear equation for benzene adsorption on the coconut-shell-based carbon. The challenge concentration was assumed to be 150 ppb, and the other variables were set as before. The results of the simulation are listed in Table 3. Again, breakthrough is shown to occur

Table 3. Variation of Exit Concentration with Bed Depth[a]

T = 4320 h (6 mo)		T = 30 h	
L (m)	C exit (ppb)	L (m)	C exit (ppb)
334.57	115	2.48	55.6
338.01	99.5	2.53	35.1
346.64	67.2	2.56	20
354.51	40.6		

[a] 150 ppb benzene challenge, 12.7 cm/s superficial velocity, 3.0-mm particle size, coconut-shell-base carbon.

quickly. Conversely, the bed length required to obtain any reasonable lifetime is prohibitively large.

This result reinforces the conclusions from the first method calculations that filter life would be extremely short, making these filters of negligible value as in-duct indoor air filters. However, it is important to note that the level at which acceptable breakthrough is set is critical to bed lifetime. In the absence of toxicological data, this level may have to be set rather arbitrarily, erring on the side of caution.

The results of this study are also supported by or verify the results reported by Daisey and Hodgson.[10] Both studies show that the small quantities of carbon used in panel and duct filters appear to have limited usefulness in controlling indoor air VOCs. Carbon devices using more complex flow schemes, involving parallel beds with regeneration, may work, but they may be uneconomical except in very large-scale applications. Carbon adsorption may be viable for major surges in concentration such as spills or releases, which push the ambient concentrations into regimes where the shape of the isotherm is more favorable and sufficient driving force exists for a reasonable uptake.

CONCLUSIONS

Air filters designed to remove particulate pollution from indoor air exhibit a minimum efficiency in the size range of importance to human health. Furnace filters provide virtually no protection from airborne particles, while industrial (ASHRAE-rate) filters do provide some protection. Nevertheless, the ASHRAE method for evaluating filter efficiency is biased toward large particles and is therefore an inadequate indicator of the protection that may be expected.

A properly designed EAC (i.e., the ESP design) can remove particles over the entire size range 0.01 to larger than 3 mm; however, the furnace filter EAC offers no advantage over the standard furnace filter in this size range. Ozone emissions from an EAC should not be of concern in a typical residential application.

Based on the measurements reported here, it appears that carbon-based sorbents at ambient conditions are not well suited for removing low levels of VOCs. Model predictions based on measured carbon capacities indicate that commercially available carbon-based duct filters are of limited utility for removing low concentrations of VOCs from indoor air.

REFERENCES

1. ASHRAE Handbook — 1983 Equipment Volume, (Atlanta: American Society of Heating, Refrigerating, and Air-Conditioning Engineers, Inc., 1983).
2. AHAM Standard Method for Measuring Performance of Portable Household Electric Cord-Connected Room Air Cleaners (AC-1). Association of Home Appliance Manufacturers (November 1987).
3. Sinclair, J.D. and Psota-Kelty, L.A., "Deposition of Airborne Sulfate, Nitrate, and Chloride Salts as It Relates to Corrosion of Electronics," *J. Electrochem. Soc.,* 137(4) (1990).
4. Castle, G.S.P., Inculet, R.I., and Burgess, K.I., "Ozone Generation in Positive Corona Electrostatic Precipitators," *I.E.E.E. Trans.,* IGA-5(4):489–496, 1969.
5. White, H.J. and Cole, W.H., "Designs and Performance Characteristics of High-velocity, High-efficiency Air Cleaning Precipitators," *JAPCA,* 10(3):239 (1960).
6. Sterling, D.A. "Volatile Organic Compounds in Indoor Air: An Overview of Sources, Concentrations, and Health Effects," in *Indoor Air and Human Health,* Gammage, R.B., Kaye, S.B., and Jacobs, V.A. (Eds.), (Chelsea, MI: Lewis Publishers, 1985), pp. 387–402.
7. Wallace, L.A., Pellizzari, E.D., Leaderer, B., Zelon, H., and Sheldon, L., "Assessing Emissions of Volatile Organic Compounds by Building Materials and Consumer Products," *Atmos. Environ.,* 21:385–393 (1987).
8. Wang, T.C., "A Study of Bioeffluents in a College Classroom," *ASHRAE Trans.,* 81(1):32–44 (1975).
9. *Vapor Phase Adsorption Handbook for the Carbon Craftsman,* (Calgon Corporation, Activated Carbon Division, 1986).
10. Daisey, J.M. and Hodgson, A.T., "Initial Efficiencies of Air Cleaners For The Removal of Nitrogen Dioxide and Volatile Organic Compounds," *Atmos. Environ.,* 23(9):1885–1892 (1989).

Part II

CHAPTER 11

Radon Overview

Jack G. Kay

Radon is a naturally occurring radioactive gas produced by decay of radium. It is ubiquitous, appearing throughout the earth's crust wherever uranium and thorium are found. It is an element that is chemically inactive in the normal environment, not very soluble in water, and is only removed by radioactive decay. In the troposphere, its chemical inertness and predictable rate of disappearance make it an ideal tracer of atmospheric dynamics and air mass movement. Its source as an atmospheric tracer is well known, i.e., the continental land mass, and its loss by decay occurs at a fixed rate, giving it a mean lifetime of 5.5 d. Man and all other living things have evolved in an environment containing radon, so inhaling radon and radon decay products is an occurrence common to all.

Radon began to be recognized as a significant component of indoor air pollution only within the last two decades. It is now understood that inhaling radon decay products provides the dominant component of natural radiation exposure for the general public.[2] Indeed, radon now is considered, along with tobacco smoke, to be the most significant pollution hazard of indoor air.

What is the nature of this health hazard, and why are we just now "discovering" it?

The answer to the second question is that, while radon has been well known as a radiation health hazard for many years, only recently have we realized that it is concentrated in the indoor environment, particularly in private homes, and that these indoor concentrations can greatly exceed the radon concentration in outdoor air.

The answer to the first question requires an understanding of what radon is. When people speak of radon, they usually are referring to the isotope, ^{222}Rn, which is produced by decay of ^{226}Ra, which is a member of the naturally occurring

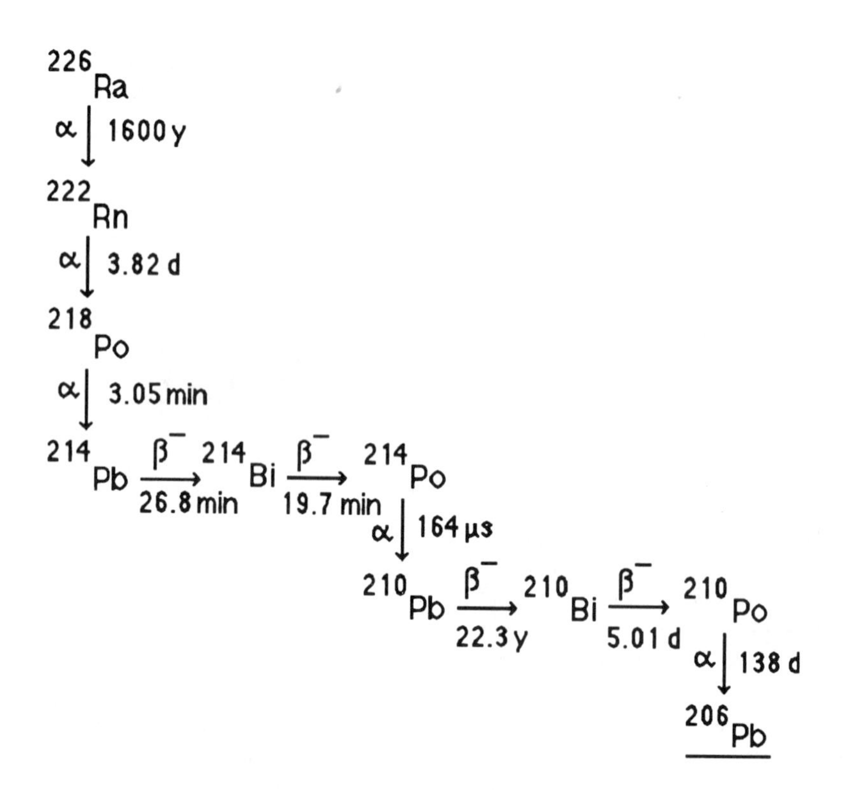

Figure 1. ^{226}Ra and progeny, part of the uranium decay series (main branch), showing decay modes and half-lives.

uranium decay chain originating with ^{238}U. This radon isotope is characterized as an alpha emitter with a half-life of 3.8 d and a mean life of 5.5 d. Its immediate decay product is ^{218}Po, half-life 3.05 min, which decays by alpha emission to produce ^{214}Pb. The decay series is depicted in Figure 1.

When ^{218}Po or one of its mass 214 progeny is inhaled, it is usually carried into the respiratory system in or on an aerosol particle. These particles most likely stick to the inner surfaces of the lungs or the bronchial passage and do not leave the body. Thus, as they undergo their natural process of radioactive decay, each particle of ionizing radiation emitted by the decaying isotopes causes radiation damage to the surrounding tissue. Since ^{210}Pb has a mean lifetime of 32 years, a long time can elapse between the time the radioisotope enters the body and its ultimate disappearance by radioactive decay. These longer-lived radon progeny can accumulate in the body over a long period of years before the results of the radiation exposure begin to be evident. Such radioisotopes often have time to move within the body to sites quite different from their original site. Radioactive lead, for example, might end up collecting in bone tissue if it is not retained in the lungs or bronchial passages.

Another radon isotope, not as well known and not as well studied as ^{222}Ra, is ^{220}Rn, commonly known as thoron since it is a member of the thorium decay series

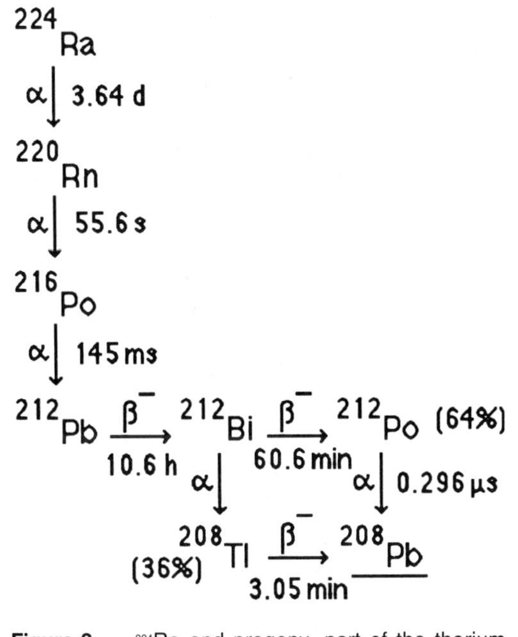

Figure 2. ^{224}Ra and progeny, part of the thorium decay series, showing decay modes and half-lives.

originating with ^{232}Th. Thoron has a half-life of only 56 s; therefore, it enters the atmosphere almost entirely by diffusion through the top few centimeters of soil.[3] Its decay by alpha particle emission produces ^{216}Po (half-life 150 msec) which immediately releases another alpha to produce ^{212}Pb (half-life 10.6 h), the longest lived radionuclide in the thoron decay chain (see Figure 2).

Both ordinary radon (^{222}Rn) and thoron can enter the indoor environment primarily through cracks in the floor and walls of the basement or any parts of the house in contact with the soil and rocks.[4] Entry can also occur via utilities, such as water, or building materials that contain traces of radium and uranium.

Radon can spread throughout a building, decaying and producing its aerosol-borne progeny everywhere. Thoron, because of its very short lifetime, is usually confined to the space where it first entered the building, but its ^{212}Pb and ^{212}Bi descendants can still be carried as aerosols into other parts of the building.

When inhaled, the mass-212 thoron daughters represent a radiation hazard to the body similar to the mass-214 radon progeny, except the delayed hazard represented by radon's mass-210 radioisotopes is non-existent in the case of thoron. This can be important, depending on how easily such progeny can diffuse through cell walls, enter the bloodstream, and be circulated to other sites within the body. There is an opportunity, especially for radon progeny, to cause radiation damage to other organs besides the lungs and bronchial passages. Thus, it might not be surprising to learn that radon exposure has been cited as the possible cause of other kinds of cancer than lung cancer.[4]

Now, let's take a closer look at the physics and chemistry that takes place when a radon or thoron atom undergoes radioactive decay. This decay is a very energetic event that results in the production of an atom different from the parent. The energy that is exchanged and the processes that occur while the new atom is adjusting to its environment can result in some unusual chemical and physical changes. When radioactive decay occurs in air, the new metal atom formed is likely to carry a positive charge and to be very reactive with any chemical species encountered. It is known that most of these radioactive products of radon decay end up as aerosol particles, i.e., submicroscopic particles that are light enough to remain airborne for long periods of time before eventually settling out on walls, floors, and other surfaces.

As a result of alpha decay, the ^{218}Po is very likely "born" with a high positive charge arising from Auger ionization and the severe perturbations caused by a sudden disruption of the nuclear charge with concomitant disturbances of the electronic orbitals of the atom. The ion also is born with lots of kinetic energy (~110 kev) because of conservation of momentum when the alpha particle is emitted.

After the ^{218}Po is born, it is a recoil nucleus that is positively charged. Thus, as a fast, heavy, charged particle, it causes ionization, excitation, and dissociation along its path in the medium in which it is produced. The recoil species regains electrons from the surroundings as it slows down near the end of its recoil path. It thus becomes thermalized, either as an ion or a neutral atom, at the end of a dense column of ion pairs and fragments with which it may react if it diffuses back in the direction that it came from. Experimental data show that about 88% of the ^{218}Po species reach the ends of their recoil paths and begin diffusing as singly charged cations. Neutralization of these cations can occur by reaction with a stray electron, by combination with a negatively charged ion, or by charge transfer involving a collision partner having a lower ionization potential than the recoil product.

Regardless of its state, whatever chemical product is formed is disrupted (destroyed) when the ^{218}Po species undergoes alpha decay to produce ^{214}Pb. The resulting lead isotope becomes another recoil ion with the same chances for neutralization or chemical combination as just described for polonium, except for minor differences in ionization potentials and recoil energies. The alpha decay process is so violent that one would not expect the final chemical state of ^{214}Pb to be at all related to the chemical state of its parent.

On the other hand, the same is not necessarily the case for ^{214}Bi produced by beta decay of ^{214}Pb. The beta decay process, while disruptive of the electronic orbitals surrounding the parent atom, is much less violent, kinetically. Recoil kinetic energies in beta decay can be zero.

The expected or predicted results for ^{214}Bi and ^{214}Po are that these species can remain on or within the aerosol particles in which they are produced. It should not be surprising, therefore, to find these radon progeny carried on larger aerosol particles than the ^{218}Po. It appears that these general expectations are borne out by experiment.

Chapters 12 and 13 in this book represent important efforts to describe the fate of ^{218}Po, both chemically and physically, in recognition of the fact that, without understanding these fundamental processes, we do not have a very good understanding of how to deal with these products of radon decay once they appear in the indoor air. The subsequent behavior of these short-lived radon progeny (and similar products of thoron decay) is dependent on their chemical and physical state.

Since the mass-214 lead and bismuth radionuclides have mean lifetimes on the order of approximately a half hour, there are numerous opportunities for things to happen to the aerosol particles bearing these isotopes before they decay. The principal hazard due to the presence of radon in indoor air is due to these aerosols that can be inhaled or ingested before the particles are deposited on the walls or other surfaces.

If these radionuclides are carried into the lungs or bronchial passages before their decay occurs, then their chemical and physical characterization is important in determining what happens in that new environment. If the particles are small enough and are not deposited in the respiratory system, they can be exhaled before they decay, much as would be expected for radon itself.

If the particles are not immediately exhaled, they will stick to the interior surfaces of the lungs or bronchial passages. Depending on whether or not the particles are water soluble they may dissolve and the radionuclide passed into the bloodstream. If the isotope survives long enough, it may be deposited somewhere else in the body before decay occurs. Regardless of where the radioisotope is when it decays, however, there is a significant risk of radiation damage to the surrounding cells or tissue. Damage to the body due to ionizing radiation can be carcinogenic; therefore, cancer is a risk whenever radionuclides of any kind decay inside the body.

If radon progeny are inhaled, there is a significant risk that they will initiate lesions that are malignant. In many instances, the radiation damage results only in destroying cells and producing mutations that are lethal to the cell (and thus not replicated) or producing nonmalignant and nonlethal mutations that the body can tolerate.

Thus, in order to address the hazards of indoor radon, it is necessary to understand how the radon progeny diffuse or are transported through the air; how they deposit on surfaces and, if inhaled or ingested, how they behave when deposited inside the body. In addition, in order to accurately assess the ultimate risk, it is necessary to understand how the ionizing radiation produced by radioactive decay of these particles interacts with the lungs or whatever tissue is near the site where decay occurs.

Chapter 14 addresses the question of sorption of radon on porous materials. The ability of radon to adsorb and then be desorbed before its decay may affect its abundance and transport in indoor air where there are lots of available surfaces for adsorption to occur. It is unlikely that radon atoms decaying on a wall will contribute to the indoor radon hazard unless some process occurs by which dust particles containing the radioactive progeny can be released from the wall surface to become airborne.

Chapter 15 discusses the problems of determining the rate of emanation of radon from a surface. This may apply to the determination of the emanating power of various soils in contact with a basement, for example, or it might apply to the release of radon from surfaces of dry wall or other construction materials indoors. The ease with which radon is released from such surfaces determines their effectiveness as potential sources of indoor radon.

Finally, Chapter 16 addresses one of the problems of experimentally determining the amount of airborne radon progeny present. Specifically, the collection and counting efficiency of wire screens is discussed. Such screens have the capability of determining not only the total number of radon progeny collected, but also the distribution of particle sizes on which the radionuclides are carried. This information is useful in predicting dry deposition rates as well as determining rates of diffusion and probabilities of sticking inside the bronchial passages and lungs.

Thus, the section of this book devoted to indoor radon addresses the important scientific questions that must be understood in order to be able to proceed with the development of effective means of control and remediation.

REFERENCES

1. Nazaroff, W.W. and Nero, A.V., (Eds.), *Radon and Its Decay Products in Indoor Air,* (New York: John Wiley & Sons, 1988), pp. ix–x, 1.
2. Schery, S.D., *J. Air Waste Manage. Assoc.,* 40:493–497 (1990).
3. Nazaroff, W.W. and Nero, A.V., (Eds.), *Radon and Its Decay Products in Indoor Air,* (New York: John Wiley & Sons, 1988), pp. ix–x, 1.
4. Martell, E.A., private communication.

CHAPTER 12

The Initial Atmospheric Behavior of Radon Decay Products

Philip K. Hopke

ABSTRACT

The chemical and physical properties of ^{218}Po immediately following its formation from ^{222}Rn decay are important in determining its behavior in indoor atmospheres and play a major part in determining its potential health effects. In 88% of the decays, a singly charged, positive ion of ^{218}Po is obtained at the end of its recoil path. The modes of neutralization, small-ion recombination, electron transfer, and electron scavenging are reviewed. In typical indoor air, the ion will be rapidly neutralized by transfer of electrons from lower ionization potential gases such as NO_2. The neutral molecule can then become incorporated in ultrafine particles formed by the radiolytic processes in the recoil path. Evidence for the formation of these particles is presented.

INTRODUCTION

The health effects of radon decay products are strongly dependent upon their aerodynamic behavior in the indoor atmosphere. Particularly for ^{218}Po, partitioning between the "unattached" state and the "attached" forms (i.e., combined with preexisting aerosol particles) has a significant impact on calculation of the dose to the lung resulting from a given airborne decay product concentration. In the dose models commonly used to relate tissue dose to airborne radioactivity concentrations,[1-3] a substantially increasing effective dose to the target tissue is predicted with decreasing particle size down to particle diameters of about 3 nm or smaller. This increased dose is due to the more effective deposition of radioactive particles in the tracheobronchial tree through diffusion as the particle size approaches the free molecular region. Small changes of size in this smallest particle size range (0.5 to 5 nm) result in large changes in the diffusion coefficient and depositional behavior, particularly in regard to the location of deposition in the tracheobronchial tree. Thus, the chemical and physical behavior of the ^{218}Po atom immediately following its formation by the alpha decay of radon must be considered if its health implications are to be fully assessed.

At the time when the alpha leaves the nucleus, there are two excess electrons around the residual polonium nucleus. It is probable that the exiting alpha causes autoionization leading to a positive ion. The recoiling polonium nucleus has a recoil energy of 117 keV with a range in air of approximately 140 μm.[4] The ion will be gas stripped to a moderate positive charge. For ^{220}Rn, the spectrum of charges on the recoiling ^{216}Po has been measured by Szucs and Delfosse[5] at reduced pressures of 10^{-5} and 1.5×10^{-4} mm Hg. They observe charges up to +9. At atmospheric pressure, even higher charge states might be anticipated. As the ion slows down toward thermal velocity at the end of the recoil path, it regains electrons such that at the end of the recoil path, 88% of the polonium atoms have a +1 charge and the remaining 12% are neutral.[6-8]

NEUTRALIZATION MECHANISMS

Background

The distribution of charge for the polonium atoms will have an important effect on their depositional behavior. The degree of neutralization will affect the apparent molecular diffusion coefficient since the ion has a diffusion coefficient about 2.5 times smaller than that of the neutral species.[7,9] The neutralization will also have an effect on the ability of electrostatic fields to collect decay products. Such collection is used in a number of radon detection systems.[10,11] Thus, understanding the processes that might interfere with collection efficiency is important to building an effective monitor.[12]

Frey et al.[13] have drawn attention to discrepancies in the literature regarding the diffusion coefficient of ^{218}Po and suggested that the degree of neutralization may play a major part in explaining these discrepancies. Another major problem in the literature was the effect of water vapor on the diffusion coefficient. Raabe[14] found increased humidity caused a decreased diffusion coefficient while Thomas and LeClare[15] found just the opposite effect in their studies. Frey et al.[13] suggested a mechanism to explain the neutralization by water vapor in which the water molecule scavenges an electron from the recoil path to form a negative ion. This scavenging produces an increased local negative ion concentration by keeping the otherwise highly mobile electrons from diffusing from the vicinity of the recoil path. They also found neutralization by 10 ppm NO_2 in nitrogen that they could not adequately explain.

Busigin et al.[16] suggested that at the end of the recoil path, polonium would react with oxygen to give an oxide species. They estimated that the polonium oxide would have an ionization potential in the 10 to 11 eV range based on analogies with lead and bismuth. This species could then extract electrons from lower ionization potential trace gases such as NO (9.25 Ev) or NO_2 (9.79 eV). Studies by Frey et al.[13] and Goldstein and Hopke[17] show that the reaction with oxygen does not itself cause neutralization. This result might be anticipated since O_2 has an ionization potential of 12.10 eV.

Goldstein and Hopke[17] narrowed the range of the possible ionization potential values for the polonium oxide species to between 10.35 eV and 10.53 eV. They could not find a gaseous compound within this range that was neither reactive nor an electron acceptor, so they could not provide a more specific value for the ionization potential. They also point out that a single water molecule cannot bind an electron.[18] Thus, they propose that the Po neutralization by water occurs because of radiolysis of the water to yield ·OH. Hydroxyl radicals are known to have an electron affinity of 1.83 eV.[19] Thus, they are excellent electron scavengers. An even better electron acceptor is NO_2 with an electron affinity of 2.32 eV. The scavenging of electrons by a high electron affinity species like ·OH or NO_2 yields an increased local concentration of negative small ions and thereby substantially enhances the small ion recombination rate. Thus, the scavenger mechanism is responsible for the neutralization results observed by Frey et al.[13] at 10 ppm NO_2 in N_2 that they could not explain. Goldstein and Hopke measured the apparent diffusion coefficient as a function of increasing NO_2 concentration from 0 to 1 ppm and found a gradually increasing value that became constant at about 700 ppb NO_2. Since 50 ppb NO_2 in O_2 yielded neutralization while 50 ppb NO_2 in N_2 yielded no apparent neutralization, they suggested that this was an indication that two different mechanisms were operating to cause neutralization of polonium ion.

In addition to these mechanisms, there is also small ion recombination.[20] However, Busigin et al.[21] report that the neutralization rates by ion recombination are of the order of seconds. Such slow rates could not explain the neutralization rates observed by Busigin et al.,[21] Frey et al.,[13] or Goldstein and Hopke.[17] Until now the rates of these neutralization processes had not been measured and the ·OH hypothesis had not been confirmed.

Studies of the rate of these various neutralization processes have recently been completed.[22,23] Using a parallel-plate diffusion chamber[24,25] in which ions can be manipulated via an applied electrostatic field, the rate for each neutralization mechanism, small ion recombination, electron transfer, and electron scavenging, have been measured. The system is illustrated schematically in Figure 1. If there is no neutralization, application of a positive or negative voltage to the bottom plate will cause the ions to move from the airstream toward the more negative plate. The top plate contains a specially designed solid state α-spectrometer with a flush face design. This plate is always kept grounded. Application of the field then increases the count in the ^{218}Po peak for a positive voltage and decreases the count for a negative voltage. The entire system has been modeled[8] so that an average time for neutralization can be extracted from the counting rate measured as a function of applied voltage. A plot showing data and the model fit for the ratio

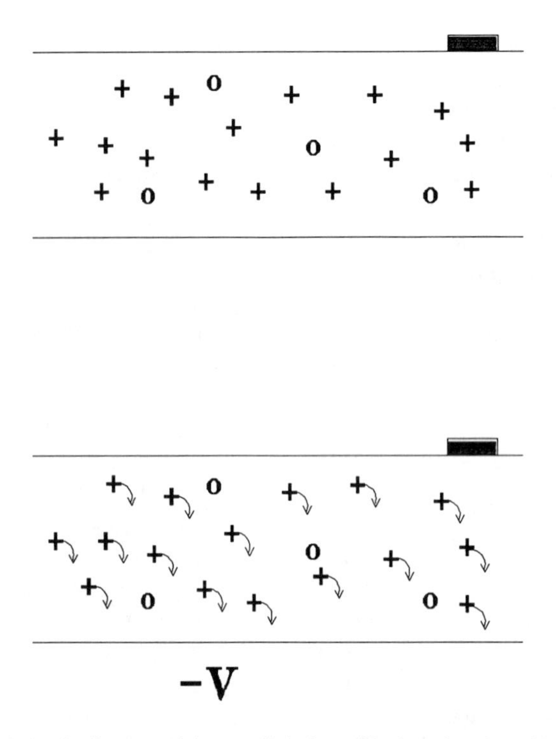

Figure 1. Schematic diagram of the parallel plate diffusion chamber showing detector in the right side of the top plate. Top plate always at ground potential. Top: without applied voltage; bottom: with voltage of −V applied to bottom plate.

of counts at a given voltage, X, to the counts obtained at zero applied field, X_o, for 250 ppb NO_2 in nitrogen is shown in Figure 2. This system can thus be used to investigate the neutralization rates for different atmospheric compositions.

Small Ion Recombination

Neutralization of $^{218}Po^+$ in nitrogen will only be via small ion recombination. Production of negative ions will be a function of radon concentration and their depletion will be related to concentrations of both positive and negative ions.

$$\frac{dC^-}{dt} = \beta[Rn] - \alpha C^+ C^- \tag{1}$$

where C^-, C^+, and [Rn] are concentrations of negative and positive ions and radon, respectively; α is the recombination coefficient; β is the negative ion production coefficient.

For steady state and equal concentrations of positive and negative ions, we then have

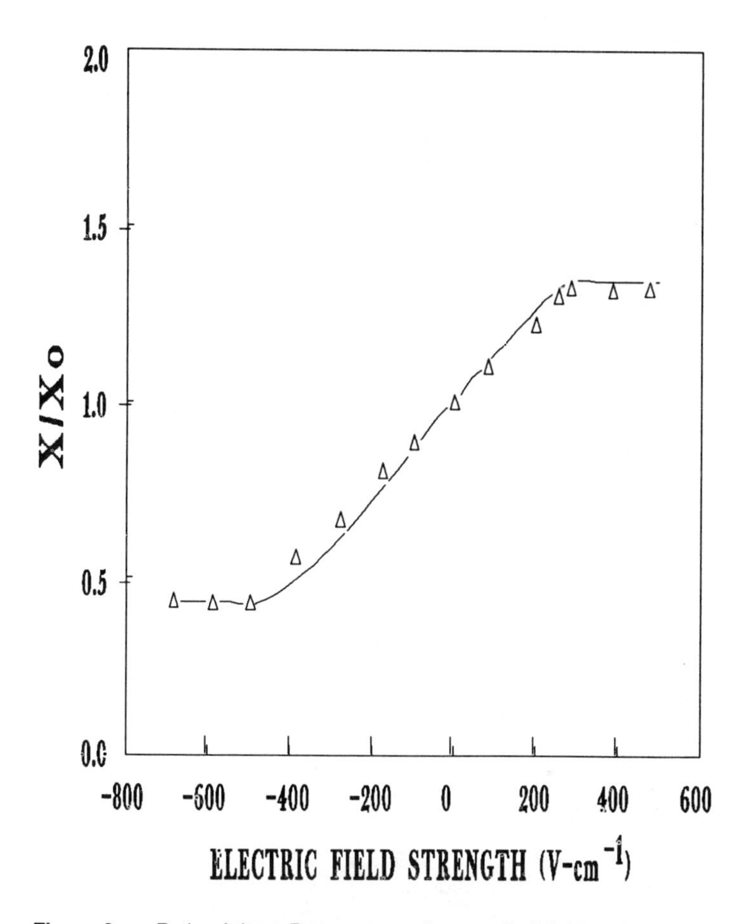

Figure 2. Ratio of the ^{218}Po counts at given applied field to the counts observed at zero field as a function of applied field strength. Points are observations for 250 ppb NO_2 in N_2; solid line is the mass balance calculation.

$$C^- = C^+ = \left(\beta[Rn]/\alpha\right)^{1/2} \tag{2}$$

Since for higher negative ion concentration there is a greater probability for polonium ions to be neutralized, the neutralization rate (R) of polonium ion is assumed to be proportional to the square root of the radon concentration, and thus can be expressed as:

$$R = a[Rn]^{1/2}+b \tag{3}$$

where a and b are constants. Figure 3, from Chu and Hopke,[23] shows that the neutralization rate as a function of the square root of the radon concentration

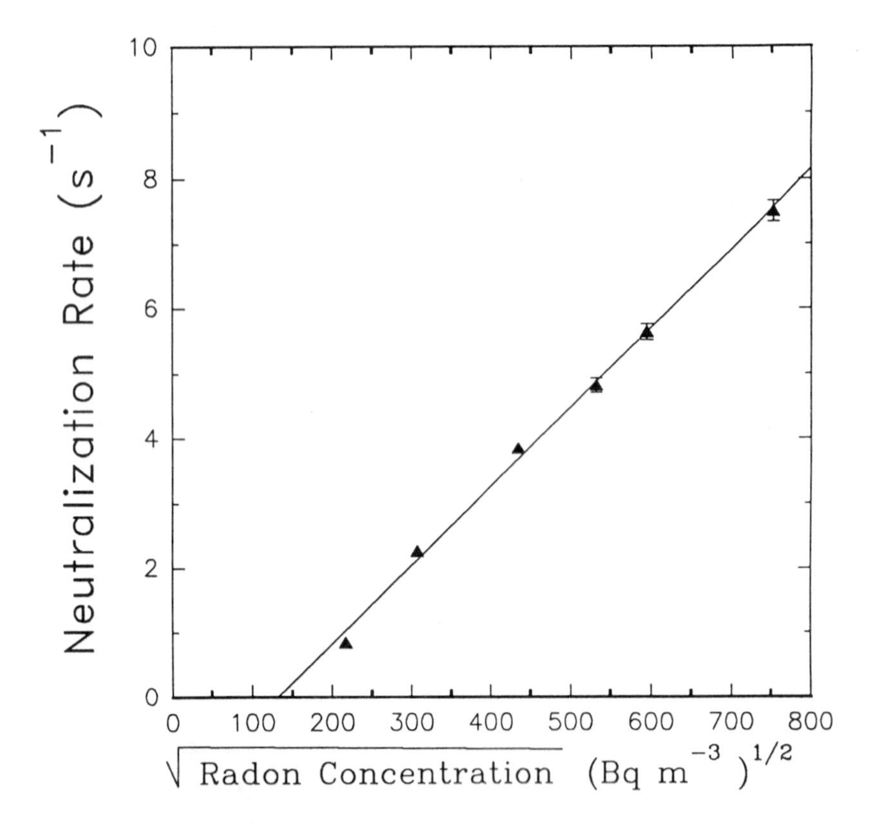

Figure 3. Neutralization rate as a function of the square root of the radon concentration in commercial grade N_2.

confirms this relationship. The intercept was attributed to diffusional losses to the walls in this relatively small chamber in agreement with the discussions of Nolan.[20]

Because this mechanism produces neutralization rates that are slow compared to the mechanisms described below, it is not very important in controlling the behavior of radon decay products. However, it does provide an opportunity to investigate more thoroughly the ion production rates in pure gases. Further studies of this mechanism have been performed by Shi[26] by studying the neutralization of $^{218}Po^{+}$ in ultrapure O_2, N_2, and Ar.

In order to evaluate the neutralization rate, the mobility of polonium ions, K, must be known. Bricard et al.[27] showed that mobility of an ion is proportional to $[(M+m)/M]^{1/2}$, where m is the mass of the gaseous molecules and M is the mass of the ion. Thus, the mobility of any ion in that gas can be obtained from the measured mobility of that ion in some known gas. Chu[8] reported the mobility of the polonium ion in nitrogen to be 1.87 ± 0.10 $cm^2/sec\text{-}V$. Consequently, the mobility of polonium ions in oxygen can be calculated as 1.89 cm^2/sec-V, and the mobility of polonium ions in argon is 1.92 cm^2/sec-V.

In order to study the neutralization rates in different radon concentrations in nitrogen, argon, and oxygen, a series of alpha counts from Po-218 under different electric field strengths in the chamber were measured. A linear regression analysis was made to determine the neutralization rate. The relationship between the neutralization rate (R) and the square root of the radon concentration ($[Rn]^{1/2}$) in N_2 can be expressed

$$R(N_2) = (0.0044 \pm 0.0005)[Rn]^{1/2} + (0.29 \pm 0.33) \tag{4}$$

and is shown in Figure 4. The regression analysis showed that the correlation coefficient is 0.961. The diffusional deposition of ions on the chamber walls is small and essentially negligible since the intercept, 0.29, in Equation 4 vanishes within the limits set by its standard error, ±0.33.

Comparison of these results and those of Chu and Hopke[23] given in Figure 3 shows that the corresponding neutralization rates are in the same order of magnitude, although neutralization rates obtained by them are higher than the present study. The difference is attributed to the different quality of gases used. As mentioned previously, the gases in this study are ultrapure grade, whereas Chu and Hopke used laboratory-grade gas (99.9% N_2, based on O_2 only). Nitrogen gas of this grade may contain water vapor concentrations as high as 32 ppm.[28] Thus, in the study by Chu and Hopke,[23] electron scavenging by a small concentration of hydroxyl radicals might occur, leading to increased neutralization and more diffusional loss to the chamber walls since the neutral polonium species has a diffusion coefficient about 2.5 larger than that for the polonium ion.[7] Thus, these current results are a better measure of small ion recombination in nitrogen.

The linear relationship between the neutralization rate (R) and the square root of the radon concentration ($[Rn]^{1/2}$) in Ar was also obtained (Figure 4) and is expressed as:

$$R(Ar) = (0.0029 \pm 0.003)[Rn]^{1/2} + (0.29 \pm 0.14) \tag{5}$$

Its correlation coefficient is 0.947. Although the intercept 0.29 is not zero considering its standard error ±0.14, it is statistically insignificant and the plateout effect of ions again is neglected. The only similar neutralization rate study in Ar were results reported by Leung and Phillips.[29] They studied the neutralization rate in Ar with humidity (minimum RH = 8 to 15%). They do not report a value for 0% RH, so a direct comparison is not possible. With a higher radon concentration and the presence of water vapor, their neutralization rates are larger as would be expected. Thus, the two studies are in general agreement.

The final relationship between the neutralization rate (R) and the square root of radon concentration ($[Rn]^{1/2}$) obtained in O_2 is also shown in Figure 4. The relationship was determined to be

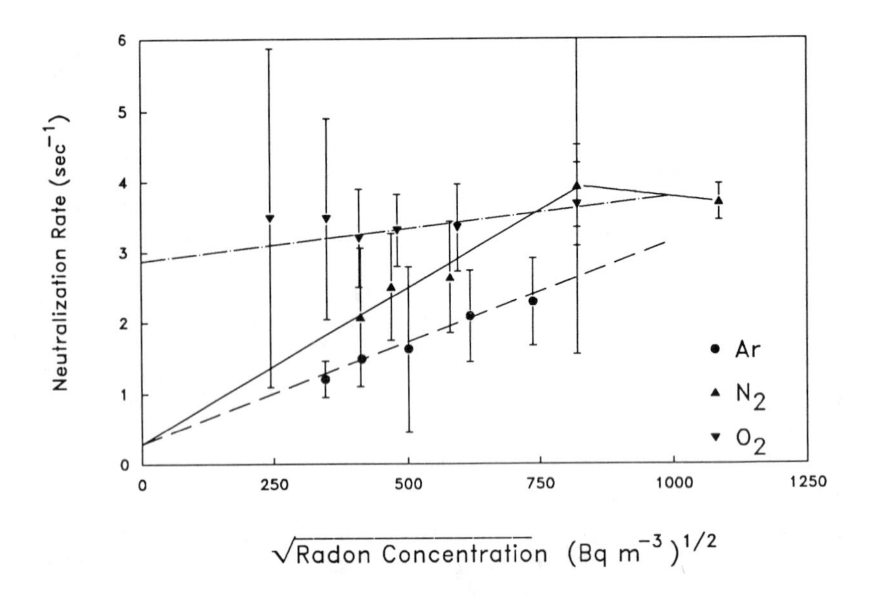

Figure 4. Neutralization rate as function of the square root of the radon concentration in high purity Ar, N_2, and O_2.

$$R(O_2) = (0.0009 \pm 0.00027)[Rn] + (2.9 \pm 0.20) \tag{6}$$

Its correlation coefficient is 0.676. The intercept 2.9 ± 0.2 in Equation 6 does not vanish. This result does not mean that there was plateout of ions on the chamber walls since the intercept is positive. Instead, it suggests that the neutralization rates in the radon concentration studied in O_2 are near or about their maximum value considering that the slope in Equation 6 is quite small. It appears that the rate has reached a plateau similar to those seen in the electron scavenging studies of Chu and Hopke.[23] When the radon concentration is sufficiently high, small negative ions will be uniformly distributed throughout the volume of gas. The polonium ion as it reaches thermal velocity then will be close enough to a negative ion so that the pair of oppositely charged ions attract each other and become neutralized. If the negative ions are only a short distance away from the Po^+, the time for neutralization is on average as short as it can be. Thus, the neutralization rate cannot increase further as the radon concentration increases.

To examine this analysis further, another set of data was measured in N_2. The neutralization rate obtained is (3.711 ± 0.258) sec^{-1} for a square root of radon concentration of 1085 (Bq m^{-3})$^{1/2}$. The largest neutralization rate previously obtained was (3.940 ± 2.924) sec^{-1} with a square root of radon concentration at 820.5 (Bq m^{-3})$^{1/2}$. Therefore, a limiting neutralization rate in N_2 also appears to exist. Further studies at high concentrations of radon in N_2 and Ar are needed to determine if this plateau behavior is real. Since neutralization rates obtained in O_2 are in the transition region between a straight line towards the origin and the neutralization rate plateau, the line of Equation 6 cannot be extrapolated to a zero radon concentration. Further results are thus needed at low radon concentration to test this hypothesis. Unfortunately, because of the large statistical counting uncertainties, the neutralization rate in lower radon concentration could not be measured.

Because of this limitation, it is not possible to ascertain whether the electron scavenging process occurred in O_2 or if the higher rate is simply due to the higher production rate of small negative ions. O_2 will capture an electron quitc quickly to form O_2^-.[30] The O_2^- is the negative counter ion to the O_2^+ formed by the original excitation process. However, because of the match of electron wave functions between the two, there will be extremely fast recombination. Thus, although there is a rapid negative ion formation, there will also be rapid removal through recombination.

A comparison of neutralization rates in O_2, N_2, and Ar made in Figure 4 showed that the neutralization rates in N_2 are lower than the corresponding ones in O_2. The N_2 rates are larger than the corresponding ones in Ar with higher ionization potential, although the differences in ionization potential are small enough to raise questions as to why there is as large a difference between these two rates as has been observed. The tendencies of the data for N_2 and O_2 indicate that there is a neutralization rate plateau for each gas once the radon concentration is sufficiently high. It will be necessary to measure the rates in argon at high ^{222}Rn to determine if similar behavior is exhibited.

Electron Transfer

The second mechanism is the electron transfer process. Neutralization is dependent on the availability of electrons on molecules with a lower ionization potential than polonium oxide. A given concentration of donor gas will yield a well defined collision rate with the ion. The effectiveness of the collision in transferring the electron will decrease as the difference in ionization potential between the polonium oxide and donor molecule decreases. The neutralization rates for NO_2 (I.P. = 9.79 eV), NH_3 (I.P. = 10.2 eV), and n-C_5H_{12} (I.P. = 10.35 eV) are shown in Figure 5 as a function of gas concentration. From these results, Chu and Hopke[23] estimate the ionization potential of the polonium oxide to be 10.44 ± 0.05 eV. Because of the rapid neutralization that will arise from this mechanism in the presence of low ionization potential trace gases, it is likely that the initially formed ^{218}PoO$^+$ will be rapidly neutralized by the organic components typically present in indoor air. Thus, it can generally be assumed that the ^{218}Po species that are not attached to preexisting indoor aerosol will be neutral. Jonassen and McLaughlin[31] have reported that less than 10% of unattached ^{218}Po are charged in a large test room. This neutralization can also interfere with radon detectors based on the electrostatic collection of the ^{218}Po$^+$.[12]

Electron Scavenging

The final mechanism, electron scavenging, was proposed by Frey et al.[13] and

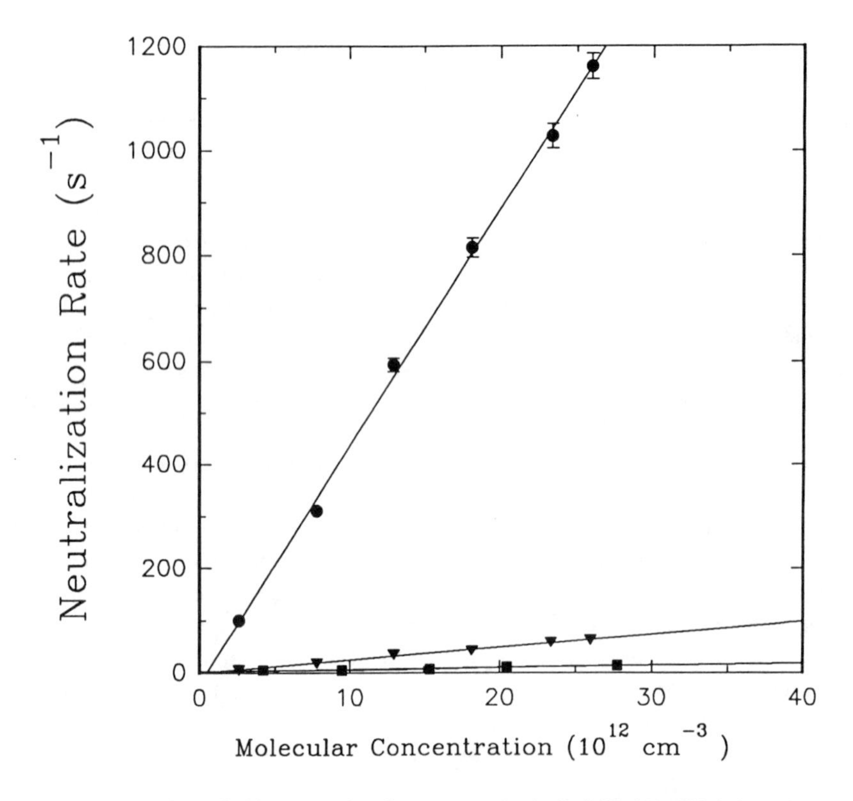

Figure 5. Neutralization rates for electron transfer in O_2 (NO_2 (●), NH_3 (▼), and n-C_5H_{12} (■)) at a radon concentration of 282.7 kBq m^{-3}.

Goldstein and Hopke[17] to explain the neutralization of polonium ions by water vapor and NO_2 in pure nitrogen where oxidation cannot occur and the electron transfer mechanism is excluded. Both Goldstein and Hopke[17] and Chu and Hopke[24] have shown an increasing diffusion coefficient with increasing NO_2 up to 700 ppb NO_2. The increasing concentration of electron capturing NO_2 leads to an increase in the local concentration of negative small ions, and thus an increased rate of neutralization compared to the case of pure nitrogen where the highly mobile electrons move rapidly from the vicinity of the polonium ion.

For the case of water vapor, Goldstein and Hopke[17] postulate the radiolysis of water to yield ·OH radicals and that it is the hydroxyl that scavenges electrons. Chu and Hopke[23] provide strong evidence in support of this hypothesis. First, if the neutralization rate for water is plotted against the water vapor concentration (Figure 6), a curved line is observed compared with the straight line behavior in NO_2 (Figure 7). However, plotting against the square root of water (Figure 8) yields a straight line. A square root behavior for ·OH production is predicted using an approach analogous to that outlined in Equations 1 and 2 for the negative small ion concentration. The fact that a straight line is obtained for the lower range of water vapor concentrations supports the hydroxyl radical hypothesis.

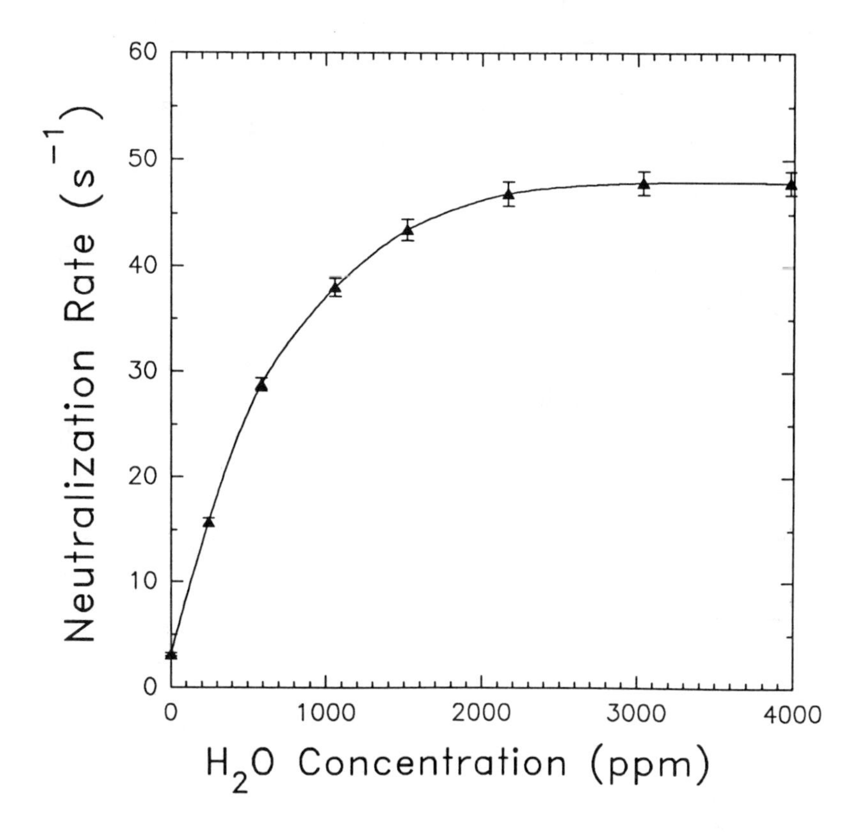

Figure 6. Neutralization rate as a function of water vapor concentration in N_2 at a radon concentration of 282.7 kBq m^{-3}.

A second test was made by adding a well known hydroxyl ion scavenger, ethanol. The resulting ethoxy ion would be expected to have a much lower electron affinity. Figure 9 shows the small ion recombination data and the two points taken with water when ethanol was added. Two values of added ethanol were studied (1.5 and 3.5 ppm). However, the measured neutralization rates were essentially identical and so they appear as the single point in Figure 9. Their proximity to the small ion recombination rate strongly supports the hydroxyl hypothesis. If it is assumed the ·OH is as effective as NO_2 in scavenging an electron and the diffusion coefficients of the two resulting ions are the same, then an ·OH concentration of 1.9×10^{-2} ppb of ·OH for each 1 ppm of water vapor can be estimated.[23] This result suggests the formation of a relatively large hydroxyl concentration in the immediate vicinity of the polonium ion at the end of its recoil path. The known importance of hydroxyl ion leads to the suggestion that reactions of hydroxyl radicals with components of indoor air could be important in modifying the chemical and physical environment of polonium ion. This behavior is discussed in the next section.

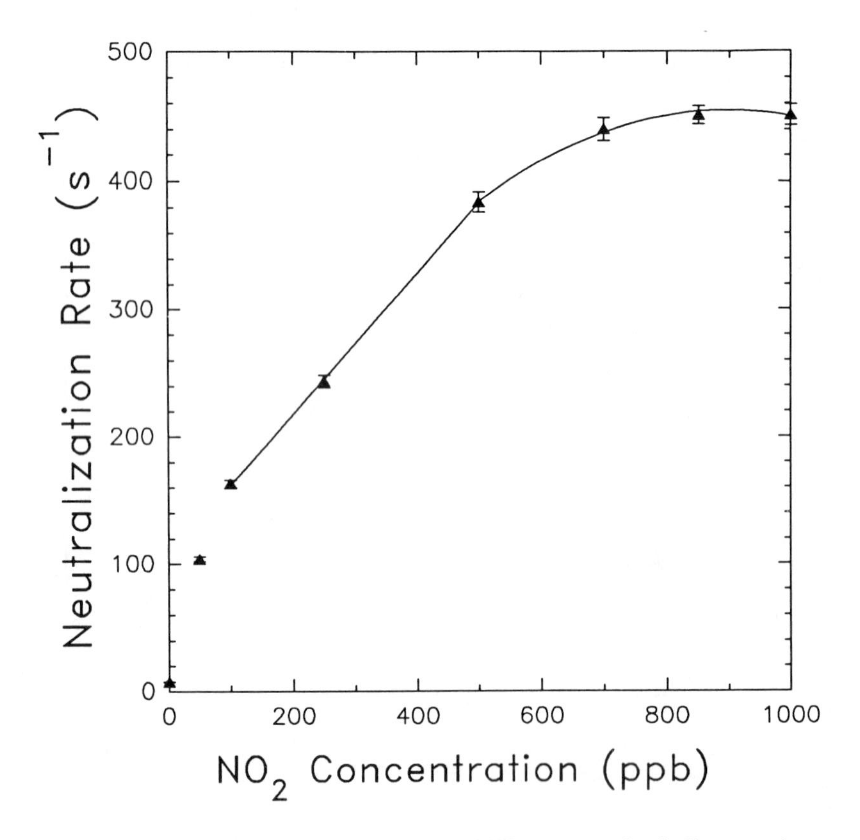

Figure 7. Neutralization rate as a function of NO_2 concentration in N_2 at a radon concentration of 282.7 kBq m^{-3}.

ULTRAFINE PARTICLE FORMATION

If ·OH radicals are being formed by the radiolysis produced by the energy deposition of both the alpha and the recoiling nucleus, these free radicals could be responsible for the production of lower volatility compounds by oxidation of trace gas precursors. These lower volatility species can nucleate around one of the many ions that exist in the vicinity of the recoil or alpha track. Depending on the availability of oxidizable compounds, nuclei can grow into particles of increasing size. At this time, only those particles to which polonium molecules are attached will be discussed. There will also be other nonradioactive particles produced in both the recoil and alpha tracks. Although a high local concentration of ·OH has been identified, the bulk average ·OH concentration is not known. Thus, the total particle production rate is also unknown until the average ·OH concentration has been measured.

It has been reported for many years that fine aerosols can be produced by

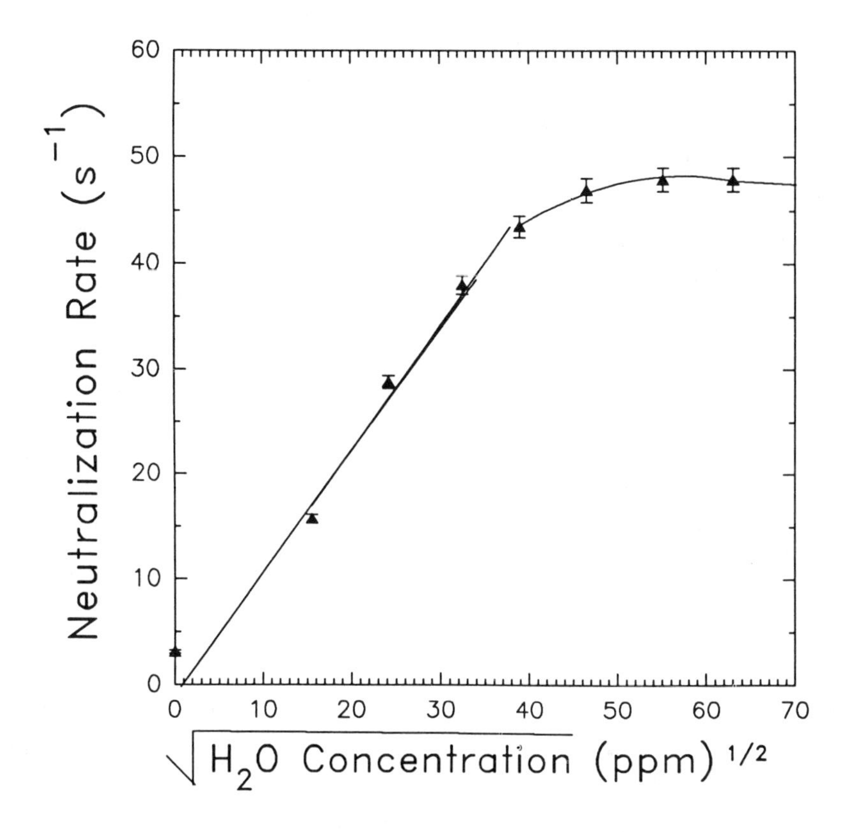

Figure 8. The neutralization rate of $^{218}Po^{+}$ in N_2 against the square root of the H_2O concentration at a radon concentration of 282.7 kBq m^{-3}.

ionizing radiation. Chamberlain et al.[32] and Megaw and Wiffen[33] present early reports of the ability of ionizing radiation to induce the formation of condensation nuclei in laboratory air. Megaw and Wiffen[33] as well as Burke and Scott[34] have suggested that there is a minimum dosage that must be deposited in a volume of air before the appearance of condensation nuclei. Similar results were obtained in our laboratory where the effects of ionization sources used to neutralize generated particles could themselves produce particles when a sufficiently high ionization density occurred in a particular volume of air.[35]

However, these results suggesting the existence of a threshold are a consequence of the method of detecting particles. The constraints of particle detectors thereby provide an operational definition of "particle". All of these previously reported studies use a condensation nuclei counter to determine the presence of particles. These devices have been found to have sharply decreasing efficiency for detecting particles less than 0.01 μm.[36,37] Thus, there is a size threshold that must be reached before a cluster of atoms becomes big enough to be detected and turns

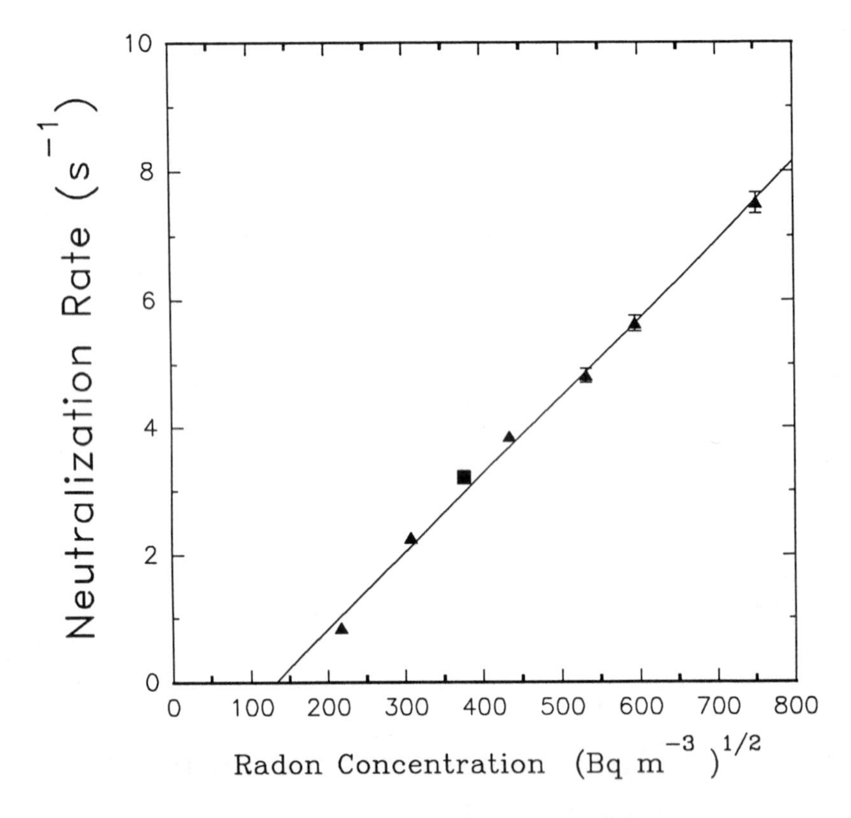

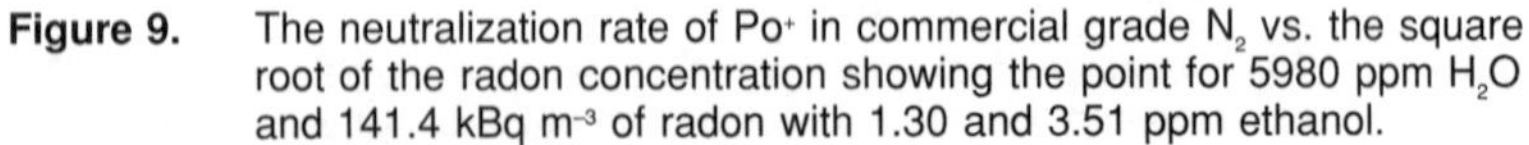

Figure 9. The neutralization rate of Po^+ in commercial grade N_2 vs. the square root of the radon concentration showing the point for 5980 ppm H_2O and 141.4 kBq m^{-3} of radon with 1.30 and 3.51 ppm ethanol.

into a "condensation nuclei". The dose threshold that had been suggested by various earlier investigators[33,34] is therefore a measurement threshold rather than one related to dose needed to produce particles.

There has been considerable study of the nature of these initially formed clusters primarily by several groups of French researchers during the 1960s and early 1970s.[27,38-43] In these experiments they typically introduced filtered, radon–laden air (usually thoron (^{220}Rn) is employed) into a large chamber, and the mobility of resulting ions is determined by drawing the ions from the chamber into either a Zeleny tube,[44] a cylinder with a negatively charged electrode positioned along the axis, or a modified Erikson spectrometer[45] as described by Bricard et al.[27] In these studies the aged air is pulled through a device that separates the ions by their electrical mobilities, and the distribution of radioactivity attached to the central wire or deposited on a nuclear emulsion is used to measure the mobility of the ions and their distribution. Blanc[39] found two major mobility groups in air in a Zeleny-type spectrometer. Fontan et al.[40] found three groups (Table 1) whose mobilities vary depending on the nature of the carrier gas used. For helium only a single group with mobility 3.3 $cm^2/s/V$ was observed.

Table 1. Mobilities in cm²/sec-V of Thoron Decay Product Ions in Several Test Gases[a]

	k_1	k_2	k_3
Nitrogen	1.80 ± 0.20	0.85 ± 0.15	0.55 ± 0.10
Oxygen	1.90 ± 0.22	0.90 ± 0.15	0.50 ± 0.14
CO_2	1.55 ± 0.23	0.75 ± 0.15	0.50 ± 0.14
Argon	1.65 ± 0.20	0.82 ± 0.15	0.40 ± 0.10

[a] Taken from Fontan et al., 1966.

In a set of experiments, Bricard et al.[42] used "air sufficiently enriched in radon or thoron to obtain an elevated density of tracks on the photographic plate" (translation of Bricard[42]) and retained the radon-laden air for about 2 min in a 2 m^3 chamber. The ion mobility spectra for both the Erikson spectrometer photographic plate and the cylindrical condenser system are quite similar, but are extremely complex with multiple peaks observed. Because of the lack of control over the trace constituent composition of the air, it is not possible to interpret the observed mobility spectra in terms of the nature of the ion clusters formed.

Fontan et al.[46] found four groups of ions with differing mobilities. Thus, there is a substantial variation in the nature of radioactive small ions observed in these studies. In most of these studies, they have used ambient laboratory air containing unknown impurity trace gases and often unknown humidity levels. Most of these studies have used plastic aging chambers that may add unknown amounts of volatile monomers to the air. Thus, it seems clear that the radiolysis is giving rise to cluster formation, but the nature of the clustering molecules cannot be inferred from these results.

In a study by Vohra et al.,[47] ethanol and an electron source were used to directly produce clusters to which radon (^{222}Rn) decay products rapidly attach. In the absence of the electron source, they find little cluster formation. This result is not surprising since small chain alcohol molecules represent extremely good free radical scavengers that could inhibit the formation of molecular clusters.[48] However, for more normal atmospheric constituents such as SO_2, ionizing radiation has been found to play a significant role in SO_2 to sulfate conversion under laboratory conditions.[47,49] In fact, Megaw and Wiffen[33] observed "some proportionality between the sulfur dioxide content of the air and the nucleus concentration produced by a given radiation dose." However they also found that the rate of nuclei formation was greater than could be explained by SO_2 alone. It has been suggested[50] that the presence of NO suppresses formation of nuclei in filtered laboratory air presumably by scavenging oxidizing free radicals.

More recent work by Madelaine and coworkers[38,51] extended the size of measurable ultrafine particles to the order of 0.003 μm and found the rapid coagulation of this ultrafine aerosol to one with a larger average diameter that would be easily observable. Martell[52] has suggested the importance of the microphysical and chemical processes that take place in the alpha track leading to the formation of nuclei from the α radiolysis. A similar process may then occur in the shorter but potentially more densely ionized region of the polonium recoil track. Perrin et al.[51] used short-lived ^{219}Rn to create their ultrafine particles. With a half-life of 3.96 s, there is a pulse of radiolysis occurring after the injection of the radon into particle-free ambient air. There is a rapid increase in observable nuclei and a longer decrease as they coagulate. The qualitative features of this process could be explained with a simple theoretical coagulation calculation.

Other studies on the radiolytic formation of particles have been made by Vohra et al.[53] and Subba Ramu and Muraleedharan.[54] In the study by Vohra and coworkers,[53] they investigated the interaction of radon with SO_2, O_3 and ethylene to form particles. Starting with moist, radon-free N_2, SO_2, and O_3 in a 4:1 N_2/O_2 mixture, they obtained 2.7×10^3 particles/cm^3. Adding ethylene to the mixture raised the level of particles to 1.6×10^4 particles/cm^3. Finally, adding 3.7 Bq m^{-3} of Rn increased the particle count to 6.6×10^4 cm^{-3}. They also observed a nonlinear increase in particles with increasing radon concentration suggesting that at low levels of radon (<3.7 Bq m^{-3}), there is a very rapid increase in ·OH production with increasing radon level. For radon in the range of 11.1 to 18.5 Bq m^{-3}, they saw a linear increase in particle production with increasing radon. At these relatively low radon concentrations, the radon helps to increase the free radical reactions by providing a direct source of ·OH from the radiolysis of water. Subba Ramu and Muraleedharan[54] have observed similar results using filtered ambient air containing unknown concentrations of SO_2, NO_x and organics to which 2.96 to 25.90 kBq m^{-3} radon is added. A proportionality of the particle production rate as measured with a condensation nuclei counter was observed with the radon concentration. It was observed that increased humidity led to an increase in the particle size of the radiolytic aerosol after a period of 200 min, suggesting some additional growth of the particles with increasing relative humidity.

Thus, it appears that there may be a continuous size range of molecular clusters up to the size at which they become particles. It appears that the rate at which the small, highly mobile clusters are formed is dose dependent, and thus at a sufficiently high dose rate, there are enough clusters that they can rapidly coagulate to activatable condensation nuclei as has been observed.

There are now new approaches that, by using the detection of the decay product activity, permit the determination of "particle" formation without a threshold. These measurement methods use the penetrability of the activity through single screens to separate particles based on their diffusion coefficients. Screen-type diffusion batteries have been used for over a decade to measure fine particle size distributions generally in conjunction with a condensation nuclei counter. Screen diffusion batteries have also been employed to measure activity size distributions. However, the commercially available screen diffusion battery uses 635 mesh screens. The penetration of 0.5 to 5 nm particles through such high mesh screens is extremely low, and there is thus limited size resolution available below 5 nm. The use of single screens of lower mesh number either in parallel or stacked in series has been described by Holub and Knutson.[55] Reineking et al.[56] and Reineking and Porstendorfer[57] have used more conventional screen batteries, but at a high enough flow as to obtain sufficient penetration to permit reasonable size determinations.

It appears that clusters in the ultrafine size range are present in indoor air. Using a high volume flow diffusion battery, studies of indoor air in houses in West Germany measuring the radon decay product activity size distribution obtained a substantial peak at a few nanometers (1 to 3 nm) and a second large accumulation mode peak in the 0.1 to 1.0 µm range.[56,57] The ^{214}Po (RaC') activity was only observed in the accumulation mode and not associated with the ultrafine particles. Knutson et al.[58] have reported similar results in indoor air with relatively low condensation nuclei counts by counting the activity attached to single screens of differing mesh size.

Only when there is a very high particle concentration is the coagulation of the ultrafine mode with the larger particle mode sufficiently rapid to remove it. It appears that under normally occurring particle concentrations, up to 50% of the ^{218}Po can be found in the ultrafine mode. Nearly all of the subsequent decay products are found to be associated with the larger particle mode.

The mechanism of particle formation has been investigated in a series of experiments conducted in the radon chamber at the Denver Research Center of the U.S. Bureau of Mines. The first experiments were described by Chu et al.[22] They find the ^{218}Po associated with an increasing sized ultrafine mode for increasing SO_2 added to air containing 37 kBq m^{-3} ^{222}Rn.

Additional experiments using similar procedures have reproduced the findings in a 2.4 m^3 chamber with fully controlled atmospheres, as shown in Figure 10. The chamber is operated under steady-state conditions as a stirred tank aerosol reactor with a flow rate of 22 lpm. These size distributions were obtained using a manual, serial graded screen array system similar to that described by Holub and Knutson.[55] The size distributions were reconstructed from the measured amounts of activity collected on each screen using the Expectation Maximization algorithm.[59] Similar results have been obtained with a parallel graded screen system.

Hopke et al.[60] has reported that the addition of NO eliminated the particle formation in accordance with the previously reported finding of Chamberlain and coworkers.[50] This result is in agreement with the ·OH hypothesis, since the reaction of NO_2 with ·OH is approximately an order of magnitude faster than the reaction with SO_2. NO_2 would be produced from the NO by reaction with either O_3 produced by radiolysis of oxygen or hydroperoxy radicals formed from the H radicals that also are produced by the radiolysis of water. Thus, the presence of NO leads to formation of gaseous nitric acid which, in contrast to sulfuric acid, will not nucleate because of its high vapor pressure.

To test if there is gaseous nitric acid being formed, another set of experiments[60] was performed where sufficient ammonia was added to titrate the gaseous nitric acid. Figure 11 shows the behavior of the ^{218}Po before adding NO to the Bureau of Mines Denver Research Center chamber, with 110 ppb NO present, and with 110 ppb NO and 10 ppm NH_3 added. These experiments were conducted at

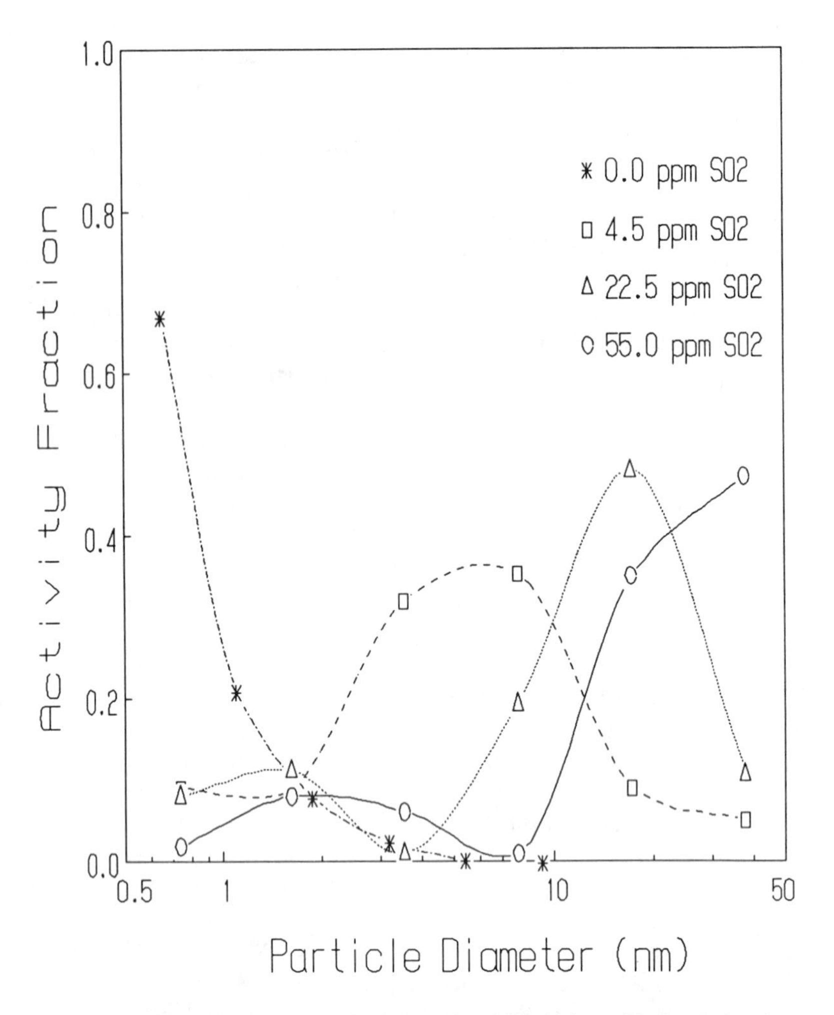

Figure 10. Size distributions in the University of Illinois (now Clarkson) chamber for the radiolytic aerosol produced in a 2.3 m³ chamber by various concentrations of SO_2 at approximately 35% relative humidity and 7400 Bq m^{-3} ^{222}Rn.

moderate relative humidities (~16%) and the small amount of growth in the background curve may be due to water clusters. Also, since the chamber uses filtered ambient air, there may be an unknown background of reactive gases in the inlet air. It appears that NO suppresses the formation of nuclei, and in the presence of a substantial excess of NH_3, there is some development of 1 to 2 nm particles. However, Figure 12 shows the size distributions for 5 ppm NO in the radon-laden air with and without added ammonia. The failure to observe a change in the activity size distribution suggests that there is insufficient NH_4NO_3 present to exceed that required for nucleation. This observation may be the result of having

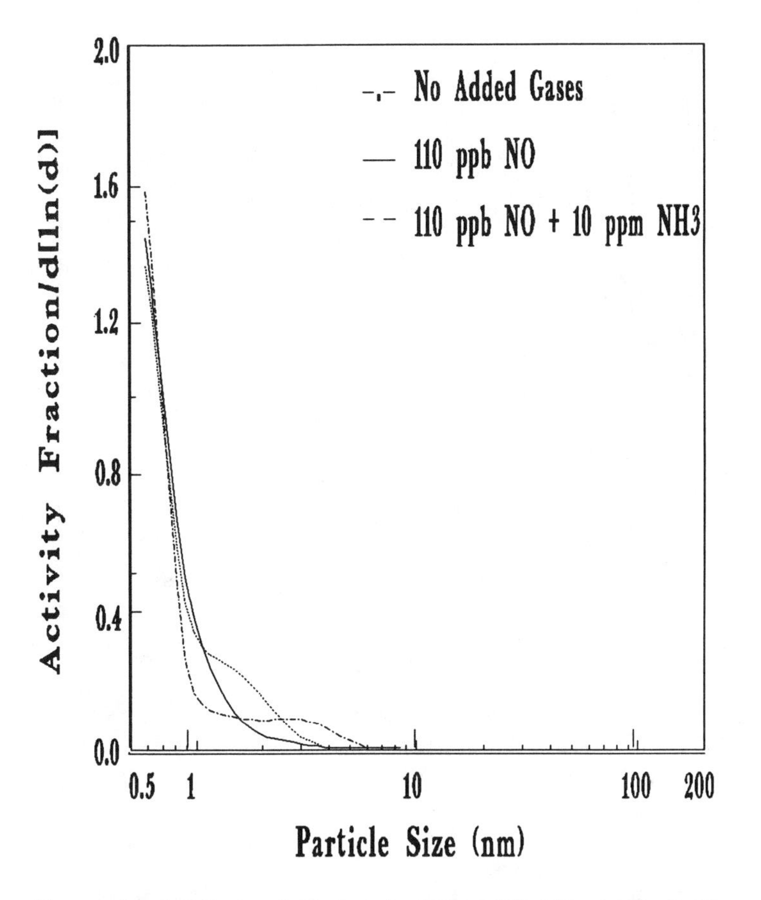

Figure 11. Activity size distributions for 100 ppb NO, 110 ppb NO with 10 ppm NH_3, and the Denver Bureau of Mines chamber background air at ~16% relative humidity.

a lower NH_3 to NO ratio. Stelson and Seinfeld[61] and Raes et al.[62] have shown that it is necessary to have a substantial excess of NH_3 to suppress the dissociation of NH_4NO_3 to NH_3 and HNO_3. Thus, the results observed are in accordance with these findings. However, if sufficient SO_2 is added, some particle growth can be observed. Further studies with greater control of gas composition are clearly needed to fully determine what reactions are occurring.

Another aspect of the radiolytic production of airborne particles by radon decay is an apparent oscillatory behavior of the size distribution of the aerosol produced in our chamber under one set of conditions. As part of the intercomparison measurements that were made in April 1988 in the 2.4 m^3 chamber in conjunction with personnel with E.M.L., Texas A & M University, U.S. Bureau of Mines, and the University of Vienna, the aerosols of various sizes giving rise to the activity size distributions in Figure 10 were generated by adding SO_2 to humidified,

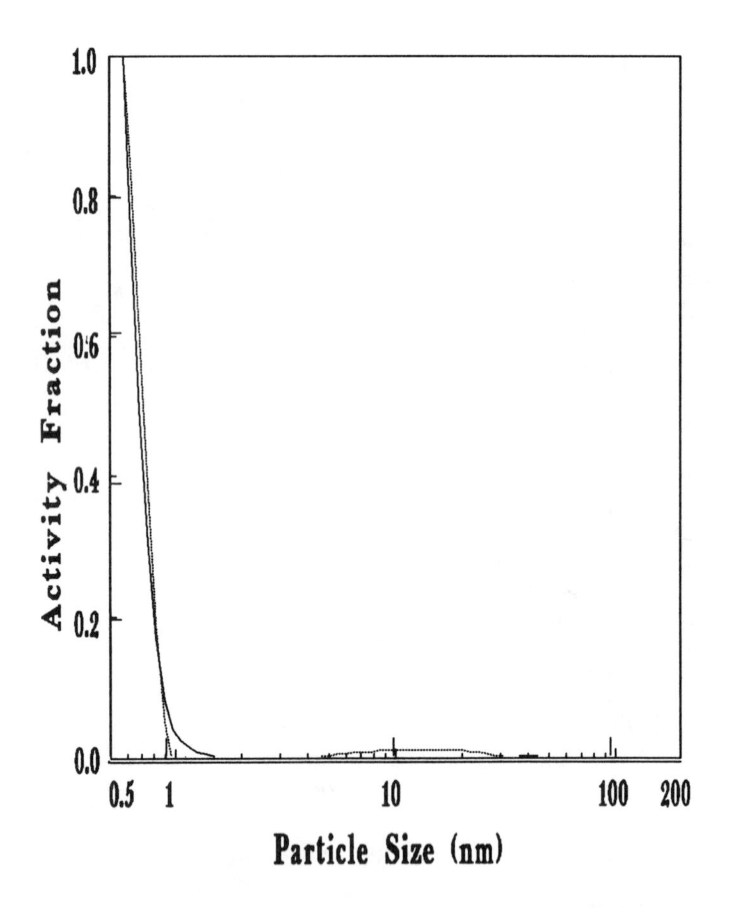

Figure 12. Activity size distribution in the Denver Bureau of Mines chamber for 5 ppm NO in 37 kBq m^{-3} ^{222}Rn with (broken line) and without (solid line) 11 ppm NH_3.

purified air containing 7400 Bq m^{-3} (200 pCi/l) radon. The activity size distributions were measured using several different graded screen array systems as described above. The steady-state particle size distributions were measured using a screen diffusion battery[63] and the University of Vienna Differential Mobility Analyzer (DMA).[64] The DMA measurements of the aerosol produced by 4.5 ppm SO_2 and 16% relative humidity are shown in Figures 13 to 18. The numbers represent the number of 8 min intervals necessary to obtain the size distribution. It can be seen that the initial distribution has its peak at about 4 to 5 nm. There is a break in time of approximately 1 h between Figure 13 and 14. In Figure 14, distribution is beginning to show a larger size mode. It appears there is coagulation of the smaller size particles into the 10 to 12 nm mode. It then becomes the dominant mode as the smallest sized particles (unobservable by the DMA) apparently coagulate directly onto these larger particles so that the 4 nm mode becomes

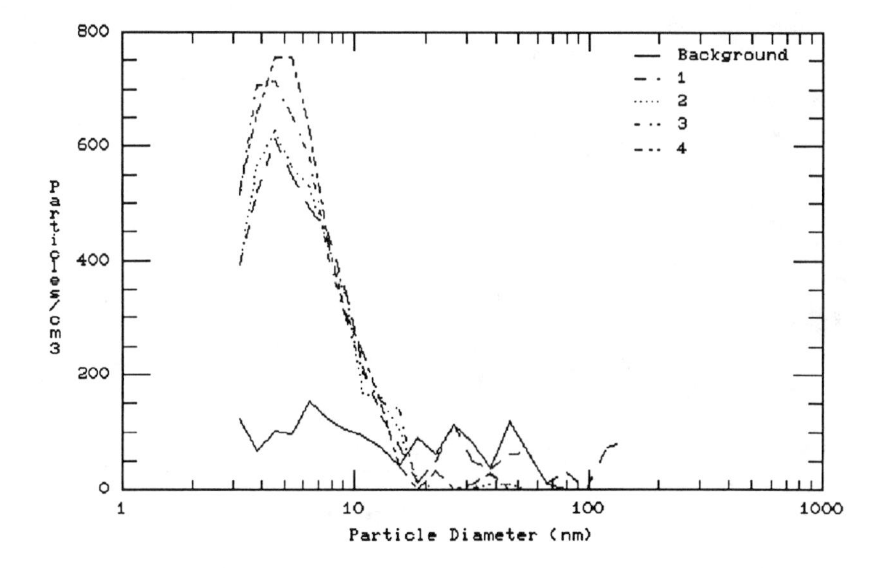

Figure 13. Size distribution of UI chamber aerosol with 4.5 ppm SO_2 and 16% R.H. during April 1988 experiments. Background spectrum and initial distributions.

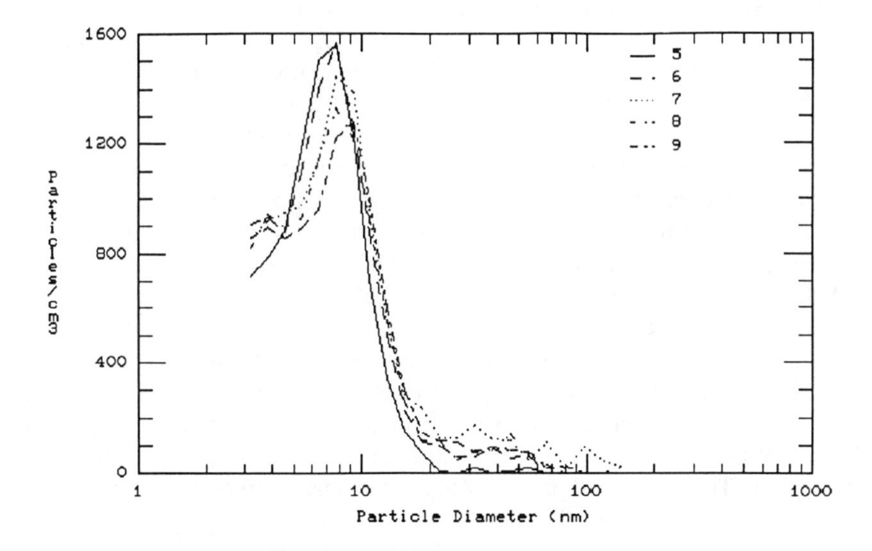

Figure 14. Size distribution of UI chamber aerosol with 4.5 ppm SO_2 and 16% R.H. during April 1988 experiments. Distributions obtained about an hour after the last of the Figure 13 distributions.

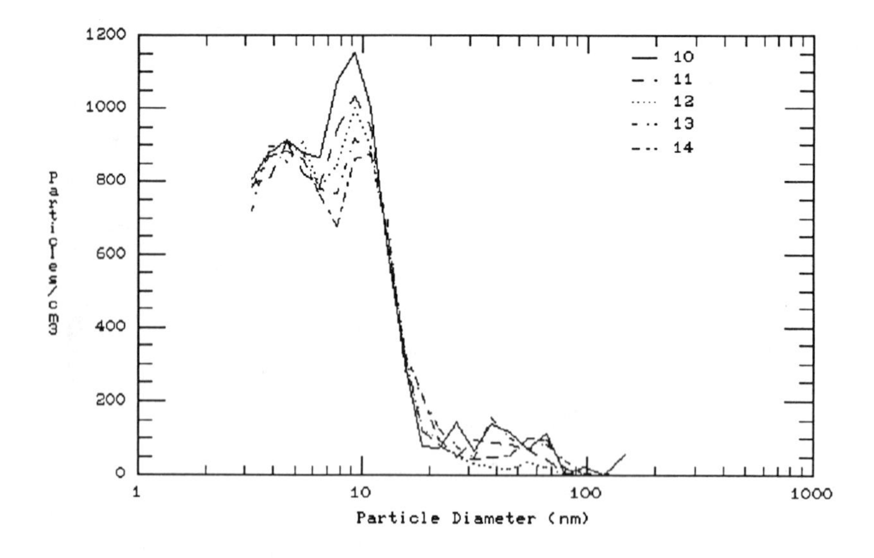

Figure 15. Size distribution of UI chamber aerosol with 4.5 ppm SO_2 and 16% R.H. during April 1988 experiments. Distributions taken at 8-min intervals.

depleted from the distribution. Subsequently, the larger mode coagulates to form fewer, larger particles to the point where there are not enough to prevent the formation of 4 to 5 nm particles, that reappear in Figure 15; by the end of the 5-h sequence (Figure 18) they once again become the dominant mode in the size distribution. Similar oscillatory behavior was observed using this same instrument in smog chamber experiments.[37] In April 1989, another set of experiments were conducted in which SO_2 and radon were used to produce an ultrafine radioactive aerosol for instrument performance intercomparison studies. These experiments help to determine whether the ^{218}Po condenses into particles because of the high concentration of H_2SO_4 in the recoil path or whether it attaches to particles formed in bulk air. At low SO_2 (2.5 ppm), no growth of the ^{218}Po was observed to sizes above ~1 nm. The activity was all in the smallest size mode (about 1 nm) and there were no countable condensation nuclei using a TSI 3020. When the concentration was raised to about 5 ppm, we immediately observed particles (25,000 cm^{-3}) and a distinct larger mode in the activity size distribution was measured. There was a definite ultrafine mode in the ^{218}Po distribution and possibly one in the ^{214}Pb distribution. Thus, these results appear to suggest that a sufficiently high relative acidity is not achieved in the vicinity of the residual polonium molecule to nucleate and it is necessary to raise the SO_2 to $SO^{=}_4$ conversion rate sufficiently high to obtain a sufficiently high, bulk average relative acidity. When that acidity is high enough, particles nucleate in large numbers throughout the chamber volume.

This hypothesis is in agreement with our previously reported, small chamber results,[60] where growth was not observed at 10 ppm SO_2 but was seen at 50 ppm SO_2. Because of greatly enhanced wall deposition of the H_2SO_4 in the small

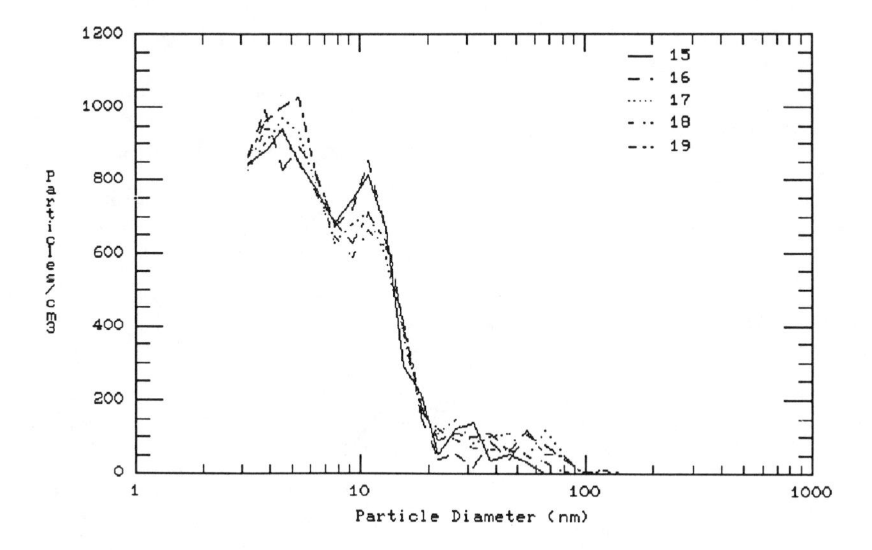

Figure 16. Size distribution of UI chamber aerosol with 4.5 ppm SO_2 and 16% R.H. during April 1988 experiments. Distributions taken at 8-min intervals.

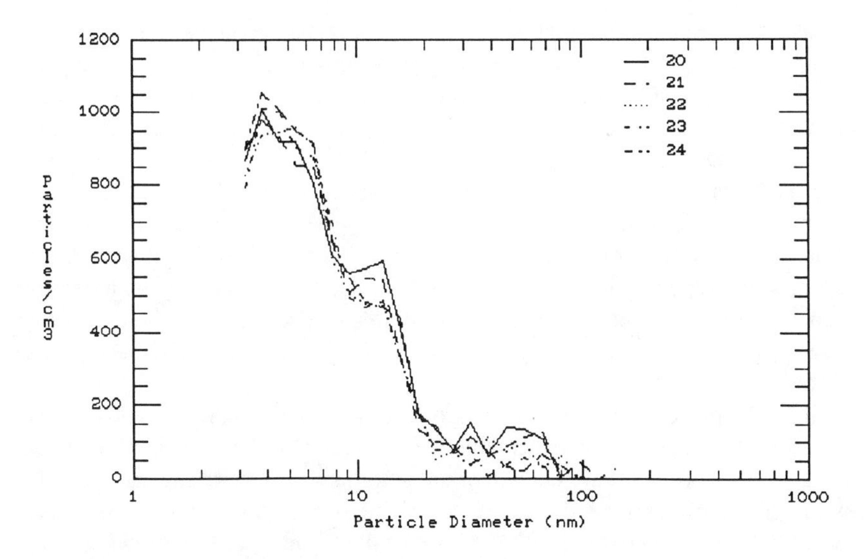

Figure 17. Size distribution of UI chamber aerosol with 4.5 ppm SO_2 and 16% R.H. during April 1988 experiments. Distributions taken at 8-min intervals.

chamber serving as an effective removal mechanism, a critical relative acidity could not be obtained without a higher SO_2 concentration necessary to increase the mass conversion rate. Thus, it appears we are not forming "particles" within the individual recoil tracks, but are adding to the acidity of the atmosphere until such time as the critical value is achieved and nucleation occurs. Thus, it is attachment

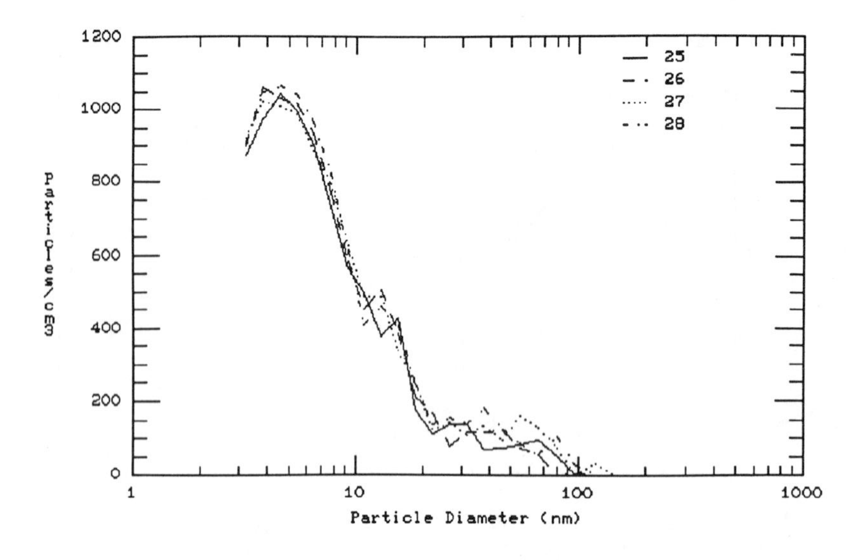

Figure 18. Size distribution of UI chamber aerosol with 4.5 ppm SO_2 and 16% R.H. during April 1988 experiments. Distributions taken at 8-min intervals.

of the decay products to the H_2SO_4 droplets rather than condensation of the acid incorporating the activity as part of the particle formation process. It also appears that there may be a threshold for forming particles. Since droplets appear not to develop until the bulk average relative acidity reaches a sufficiently high value, there must be sufficient conversion of SO_2 to $SO^=_4$ to achieve the concentration required for nucleation. It is likely that the previously determined dose "threshold" determinations for radiolytic nuclei formation were dominated by instrumental limits; we need to carefully define the potential existence of limiting values of radon, water vapor, and reactive trace gas that define the threshold boundary for particle formation. In addition, such a threshold will also be influenced by whether the $SO^=_4$ is in the form of H_2SO_4 or has been partially neutralized by the presence of NH_3.

Thus, the presence of airborne compounds that react with hydroxyl radicals to form lower volatility compounds will cause a change in the size distribution and the diffusivity of the ultrafine mode of the polonium activity size distribution. An excellent framework for modeling this ultrafine particle formation has been presented by Raes and co-workers.[65,66] More detailed studies of the mechanism and its dependence on reactant and water vapor concentration are in progress. It does appear likely that in most indoor atmospheres there will be sufficient reactive gases to cause the "unattached" fraction to be a distribution of size with physical properties such as diffusion coefficients that can vary by over an order of magnitude. To characterize the airborne radioactivity fully in terms of its aerodynamic behavior and therefore its deposition and potential dose, it will be necessary to characterize the whole activity size distribution including the ultrafine particle mode.

CONCLUSIONS

The recent results summarized above begin to provide a basis for understanding many of the reports in the literature on observed variations in the diffusion coefficient of ^{218}Po. The combination of neutralization and ultrafine particle formation can produce a wide range of possible diffusivities depending on the exact composition of the atmosphere under study. However, from the rate data presented here, it seems that for most indoor atmospheres there will be a rapid neutralization of the polonium oxide ion. Thus, it appears that the apparent diffusivity of the polonium will be governed by the size of cluster formed, which in turn will depend on the concentrations of reactive gases present in the air. With improved experimental measurements coupled with better theoretical models, it is likely that any particular situation will be able to be quantitatively understood.

ACKNOWLEDGMENTS

Much of the work reported here was performed at the University of Illinois with the support of the U.S. Department of Energy under Contract DE AC02-83ER60186 and DE FG02-87ER60546 and at Clarkson University under Contracts DE FG02-89ER60860 and DE FG02-89ER60876. I would like to thank the many students whose work is described in this chapter for their efforts in these studies.

REFERENCES

1. Jacobi, W. and Eisfeld, K., "Dose to Tissues and Effective Dose Equivalent by Inhalation of Radon-222, Radon-220, and Their Short-Lived Daughters," Gesellshaft fur Strahlen- and Umweltforschung GSH Report S-626 (Munich, West Germany, 1980).
2. James, A.C., Greenhalgh, J.R., and Birchall, A., "A Dosimetric Model for Tissues of the Human Respiratory Tract at Risk from Inhaled Radon and Thoron Daughters." in *Radiation Production, A Systematic Approach to Safety, Proc. 5th Congress IRPA,* Jerusalem: March 1980. Vol. 2 (Oxford: Pergamon Press, 1980), pp. 1045–1048.
3. Harley, N.H. and Pasternack, B., "Environmental Radon Daughter Alpha Dose Factors in a Five-Lobed Human Lung," *Health Phys.,* 42:789–799 (1982).
4. Lind, S.C., *Radiation Chemistry of Gases*, (New York: Reinhold Publishing Corp., 1961).
5. Szucs, S. and Delfosse, J.M., "Charge Spectrum of Recoiling ^{218}Po in the α Decay of ^{220}Rn," *Phys. Rev. Lett.,* 15:163–165 (1965).
6. Wellisch, E.M., "The Distribution of the Active Deposition of Radium in an Electric Field—II," *Philos. Mag.,* 6(xxvi):623–635 (1913).

7. Porstendorfer, J. and Mercer, T.T., "Influence of Electric Charge and Humidity Upon the Diffusion Coefficient of Radon Decay Products," *Health Phys.*, 15:191–199 (1979).
8. Chu, K.D., "The Kinetics of Neutralization of Po-218," Ph.D. thesis, (University of Illinois, Urbana, 1987).
9. Porstendorfer, J., "The Diffusion Coefficients and the Mean Free Paths of the Neutral and Charged Radon Decay Products in Air," *Z. Physik,* 213:384–396 (1968).
10. Wrenn, M.E., Spitz, H., and Cohen, N., "Design of a Continuous Digital-Output Environmental Radon Monitor," *IEEE Trans. Nucl. Sci.,* NS-22:645–648 (1975).
11. Porstendorfer, J., Wicke, A., and Schraub, A., in *Natural Radiation Environment III,* Gesell, T. and Lowder, W.M., (Eds.), 2:1293–1307 (1980).
12. Hopke, P.K., "On the Use of Electrostatic Collection of Radon Decay Products for Measuring Radon," *Health Phys.,* 57:39–42 (1989).
13. Frey, G., Hopke, P.K., and Stukel, J.J., "Effects of Trace Gases and Water Vapor on the Diffusion Coefficient of Polonium-218," *Science,* 211:480–481 (1981).
14. Raabe, O.G., "Measurement of the Diffusion Coefficient of Radium A," *Nature (London),* 217:1143–1145 (1968).
15. Thomas, J.W. and LeClare, P.C., "A Study of the Two-Filter Method for Radon-222," *Health Phys.,* 18:113–122 (1970).
16. Busigin, A., van der Vooren, A.W., Babcock, J.C., and Phillips, C.R., "The Nature of Unattached RaA (^{218}Po) Particles," *Health Physics,* 40:333–343 (1981a).
17. Goldstein, S.D. and Hopke, P.K., "Environmental Neutralization of RaA (^{218}Po)," *Environ. Sci. Technol.,* 19:146–150 (1985).
18. Jordan, K.D. and Wendoloski, J.J., "On the Existence of Negative Ions of Nonionic Polar Molecules: Studies of HF, H_2O^- HCN^-, $(HF)^{-2}$, H_3NO^-, and CH_3CN^-," *Health Phys.,* 21:145–54 (1977).
19. Drzaic, P.S., Marks, J., and Brauman, J.I., *Electron Photodetachment from Gas Phase Molecular Anions, in Gas Phase Ion Chemistry, Vol. 3,* Bowers, M.R., (Ed.), (New York: Academic Press, 1984), pp. 167–211.
20. Nolan, P.J., "The Recombination Law for Weak Ionization," *Proc. R.I.A.,* 49(A):67–90 (1943).
21. Busigin, C., Busigin, A., and Phillips, C.R., "The Chemical Fate of ^{218}Po in Air, Golden, CO," in *Radiation Hazards in Mining: Control, Measurement and Medical Aspects*" Gomez, M., (Ed.), (Kingsport, TN: Kingsport Press, 1981b), pp. 1043–1047.
22. Chu, K.D., Hopke, P.K., Knutson, E.O., Tu, K.W., and Holub, R.F., "The Induction of an Ultrafine Aerosol by Radon Radiolysis," in *Radon and Its Decay Products: Occurrence, Properties and Health Effects,* Hopke, P.K., (Ed.), (Washington, DC: American Chemical Society, 1987), pp. 365–376.
23. Chu, K.D. and Hopke, P.K., "Neutralization Kinetics for Polonium-218," *Environ. Sci. Technol.,* 22:711–717 (1988).
24. Chu, K.D. and Hopke, P.K., "Continuous Monitoring Method of the Neutralization Phenomena of Polonium-218," Paper No. 85-85.5, (Pittsburgh: Air Pollution Control Association, 1985).
25. Chu, K.D. and Hopke, P.K., "Continuous Monitoring Method in the Neutralization Rate Analyses of Polonium-218," Paper No. 86-81.10, (Pittsburgh: Air Pollution Control Association, 1986).

26. Shi, B., "The Study of Neutralization of Po-218 Ions by Small Ion Recombination in O_2, Ar, and N_2," M.S. thesis, (University of Illinois, Urbana, May 1989).
27. Bricard, J., Girod, P., and Pradel, J., "Spectre de Mobilite des Petits Ions Radioactifs de l'Air," *C.R. Acad. Sci. Paris*, Groupe 6(260):6587–6590 (1965).
28. *CGA Handbook of Compressed Gases*, 3rd Edition, (New York: Van Nostrand Reinhold, 1989).
29. Leung, H.M.-Y. and Phillips, C.R., "The Electrical and Diffusive Properties of Unattached ^{218}Po in Argon Gas," *Radiat. Prot. Dosimetry*, 18:3–11 (1987).
30. Korff, S.E., *Electron and Nuclear Counters: Theory and Use*, Second Edition, (New York: D. Van Nostrand Company, 1955), pp. 119–123.
31. Jonassen, N. and McLaughlin, J.P., "The Reduction of Indoor Air Concentrations of Radon Daughters Without the Use of Ventilation," *Sci. Total Environ.*, 45:485–492 (1985).
32. Chamberlain, A.C., Megaw, W.J., and Wiffen, R.D., "Role of Condensation Nuclei as Carriers of Radioactive Particles," *Geofis. Pura Appl.*, 36:233–242 (1957).
33. Megaw, W.J. and Wiffen, R.D., "The Generation of Condensation Nuclei by Ionizing Radiation," *Geof. Pura Appl.*, 50:118–128 (1961).
34. Burke, T.P. and Scott, J.A., "The Production of Condensation Nuclei by Alpha Radiation," *Proc. R. Irish Acad.*, 73:151–158 (1973).
35. Leong, K.H., Hopke, P.K., Stukel, J.J., and Wang, H.C., "Radiolytic Condensation Nuclei in Aerosol Neutralizers," *J. Aerosol Sci.*, 14:23–27 (1983).
36. Agarwal, J.K. and Sem, G.J., "Continuous Flow, Single-Particle Counting Condensation Nucleus Counter," *J. Aerosol Sci.*, 11:343–357 (1980).
37. Wang, H.C., Leong, K.H., Stukel, J.J., and Hopke, P.K., "Calibration of an Autoranging Condensation Nuclei Counter," *Atmos. Environ.*, 16:2999–3001 (1982).
38. Madelaine, G.J., Perrin, M.L., and Renoux, A., "Formation and Evolution of Ultrafine Particles Produced b6 Radiolysis and Photolysis," *J. Geophys. Res.*, 85:7471–7474 (1980).
39. Blanc, D., Fontan, J., and Juan, M., "Sur les Mobilites d'Ions Radioactifs Naturels Obtenus en Atmosphere Filtree," *C.R. Acad. Sci. Paris*, 257:2099–2100 (1963).
40. Fontan, J., Billard, F., Blanc, D., Bricard, J., Huertas, M.L., and Marty, A.M., "Etude de la Mobilite des Ions Radioactifs Formes sur les Atomes de Redul Provenant de la Disintegration du Thoron Dans l'Air et Differents Gaz," *C.R. Acad. Sci. Paris Ser. B*, 262:1315–1317 (1966).
41. Billard, F., Bricard, J., Cabane, M., and Madelaine, G., "Etude des Noyaux de Condensation Qui se Formeut Dans l'Obsurite et Sous l'Action de la Lumiere Dans l'Air Exempt d'Aerosols," *C.R. Acad. Sci. Paris Ser. B*, 265:1173–1176 (1967).
42. Bricard, J., Billard, F., Blanc, D., Cabane, M., and Fontan, J., "Structure Detaillee du Spectre de Mobilite des Petits Ions Radioactifs Dans l'Air," *C.R. Acad. Sci. Paris Ser. B*, 263:761–764 (1966).
43. Bricard, J., Billard, F., and Madelaine, G., "Formation and Evolution of Nuclei of Condensation that Appear in Air Initially Free of Aerosols," *J. Geophys. Res.*, 73:4487–4496 (1968).
44. Zeleny, J. "The Velocity of the Ions Produced in Gases by Rontgen Rays," *Philos. Trans. R. Soc. (London)*, 195(A):193–234 (1901).
45. Erikson, H.A., "On the Nature of the Negative and Positive Ions in Air, Oxygen, and Nitrogen," *Phys. Rev.*, 20(2):117–126 (1922).
46. Fontan, J., Blanc, D., Hubertas, M.L., and Mart, A.M., "Mesure de la Mobilite et du

Coefficient de Diffusion des Particules Radioactives," in *Planetary Electrodynamics,* Coroniti, S.C., and Hughes, J., (Eds.), (New York: Gordon and Breach Science Publishers, 1969), pp. 257–267.
47. Vohra, K.G., Subba Ramu, M.C., and Rao, A.M.M., "A Study of the Mechanism of Formation of Radon Daughter Aerosols," *Tellus,* 18:672–678 (1966).
48. Gusten, H., Filby, W.G., and Schoaf, S., "Prediction of Hydroxyl Radical Reaction Rates with Organic Compounds in the Gas Phase," *Atmos. Environ.,* 15:1763–1765 (1981).
49. Vohra, K.G., "Gas-to-Particle Conversion in the Atmospheric Environment by Radiation-Induced and Photochemical Reactions, in Radiation Research," Nygaard, D.F., Adler, H.I., and Sinclair, W.K., (Eds.), (New York: Academic Press, 1975), pp. 1314–1325.
50. Chamberlain, A.C., Heard, M.J., Penkett, S.A., and Wells, A.C., "Suppression of Radiolytic Nuclei in Air by Nitric Oxide," *Health Phys.,* 37:706–707 (1969).
51. Perrin, M.L., Margne, J.P., and Madelaine, G.J., "Etude Experimentale et Theorique de l'Evolution d'un Aerosol de Dimension Inferieure a 0.02 μm," *J. Aerosol Sci.,* 9:429–433 (1978).
52. Martell, E.A., "Radioactive Emanations and the Formation and Composition of Aerosols," in *Proc. Garmisch-Partenkirchen Symposium on Radiation in the Atmosphere 44,* (Princeton, NJ: Science Press, 1977).
53. Vohra, K.G., Subba Ramu, M.C., and Muraleedharan, T.S., "An Experimental Study of the Role of Radon and its Daughter Products in the Conversion of Sulphur Dioxide into Aerosol Particles in the Atmosphere," *Atmos. Environ.,* 18:1653–1656 (1984).
54. Subba Ramu, M.C. and Muraleedharan, T.S., "Formation of Aerosols in Irradiated Atmospheric Air," *J. Aerosol Sci.,* 17:745–751 (1986).
55. Holub, R.F. and Knutson, E.O., "Measurement of Po-218 Diffusion Coefficient Spectra Using Multiple Wire Screens," in *Radon and Its Decay Products: Occurrence, Properties and Health Effects,* Hopke, P.K., (Ed.), (Washington, DC: American Chemical Society, 1987) pp. 365–376.
56. Reineking, A., Becker, K.H., and Porstendorfer, J., "Measurements of the Unattached Fractions of Radon Daughters in Houses," *Sci. Total Environ.,* 45:261–270 (1985).
57. Reineking, A. and Porstendorfer, J., "High-Volume Screen Diffusion Batteries and α-Spectroscopy for Measurement of the Radon Daughter Activity Size Distributions in the Environment," *J. Aerosol Sci.,* 17:873–879 (1986).
58. Knutson, E.O., George, A.C., Hinchliffe, L. and Sextro, R., "Single Screen and Screen Diffusion Battery Method for Measuring Radon Progeny Size Distributions, 1-500 nm," (Presented to the 1985 Annual Meeting of the American Association for Aerosol Research, Albuquerque, NM, 1985).
59. Maher, E.F. and Laird, N.M., "Em Algorithm Reconstruction of Particle Size Distributions From Diffusion Battery Data," *J. Aerosol Sci.,* 16:557–570 (1985).
60. Hopke, P.K., Chu, K.D., Kulju, L., and Holub, R.F., "Formation of Ultrafine Particles by Radon Radiolysis," (Presented to the 1987 Annual Meeting of the American Association for Aerosol Research, Seattle, 1987).
61. Stelson, A.W. and Seinfeld, J.H., "Relative Humidity and Temperature Dependence of the Ammonium Nitrate Dissociation Constant," *Atmos. Environ.,* 16:983–992 (1982).

62. Raes, F., Kodas, T.T., and Friedlander, S.K., "Aerosol Formation in Chemically Reacting Gases: Application to Ammonium Nitrate Aerosols," (Presented to the AAAR '87 Annual Meeting, Seattle, 1987a).
63. Ramamurthi, M., "The Detection and Measurement of the Activity Size Distributions (d_p > 0.5 nm) Associated With Radon Decay Products in Indoor Air," Ph.D. thesis, (University of Illinois, Urbana, 1989).
64. Winklemayr, W., "Untersuchen des Ultrafeinen Aerosols in der Urbanen Atmosphäre von Wien," Ph.D. thesis, (University of Vienna, 1987).
65. Raes, F., "Description of the Properties of Unattached ^{218}Po and ^{212}Pb Particles by Means of the Classical Theory of Cluster Formation," *Health Phys.*, 49:1177–1187 (1985).
66. Raes, F., Janssens, A., and Vanmarcke, H., "A Model for Size Distributions of Radon Decay Products in Realistic Environments," in *Radon and Its Decay Products: Occurrence, Properties and Health Effects,* Hopke, P.K., (Ed.), (Washington, DC: American Chemical Society, 1987b), pp. 324–339.

CHAPTER 13

Properties of Unattached ^{218}Po in Gas Systems Containing Trace Amounts of Hydrocarbons (Hexane, Ethylene, or Methane)

William W.-C. Chan and Colin R. Phillips

ABSTRACT

The electrostatic and diffusion properties of unattached ^{218}Po were investigated in an inert gas system containing trace amounts of hydrocarbons. Hexane (ionization potential, I.P., = 10.18 eV), ethylene (I.P. = 10.5 eV), and methane (I.P. = 12.6 eV) were used as the trace gases. The diffusion coefficient of neutral unattached ^{218}Po was measured at various relative humidities. The low diffusion coefficient found at high humidity was attributed to the formation of clusters. The fraction of ^{218}Po having a positive charge at a gas age of about 0.1 sec was determined. Based on the time-of-flight method, a pulse-width modulated ion mobility analyzer was used to determine the mobility of the ^{218}Po. For a singly charged ^{218}Po ion, the diffusion coefficient and mobility data obtained were consistent with the Einstein equation. The neutralization rate constant of ^{218}Po ions was investigated at various relative humidities and radon concentrations. Three neutralization mechanisms were inferred. The neutralization rate constant was found to be proportional to a fractional power exponent of the radon concentration, consistent with the small ion recombination mechanism. The charge transfer mechanism involving hydrated polonium species and hexane (or ethylene) was considered responsible for lowering the power dependence of the neutralization rate constant on radon concentration from 0.5 to 0.4 (or 0.45). Charge transfer does not occur in the methane-nitrogen gas system because of the high ionization potential of methane. The power dependence was found to increase with increasing ionization potential of trace gas and to level off at 0.5 as the ionization potential becomes too high to allow charge transfer. The ionization potential of hydrated polonium species was estimated to be about 11 eV. The neutralization rate constant was found to increase with relative humidity, suggesting involvement of the hydrated polonium species in charge neutralization.

INTRODUCTION

It is well established that exposure to radon and its progeny results in an increased risk of lung cancer.[1] High levels of airborne radioactivity have been reported not only in uranium mines, but also in houses.[2] The health hazards associated with exposure to indoor radon and its progeny have been the subject of recent review. Nero concluded that "the present exposure of the general public to radon progeny may account for a substantial number of lung cancers in the U.S." and calculated the risk as an annual incidence of 10 to 100 per million for an annual exposure of 0.2 WLM.[2] Application of advanced energy conservation technology in houses would further increase the risk. Based on recent surveys of indoor air, the National Council on Radiation Protection increased its estimate of dose to the bronchial epithelium from 450 mrem in 1975 to 3000 mrem in 1984.[3,4] Important public health objectives are to identify the level of risks associated with exposure to radon and its progeny, to determine the dose to the critical cells of the respiratory tract, to identify the particle deposition patterns and removal mechanisms, and to develop new and improved measurement methodologies for confined environments. These objectives can be achieved only after the fundamental electrostatic and diffusion properties of radon progeny are understood.

^{218}Po is the first short-lived progeny of the radon (^{222}Rn) decay chain. It is formed from the alpha-decay of ^{222}Rn. The recoiling ^{218}Po atom has an energy of 110 keV.[5] At the end of the recoil path, ^{218}Po atom is fully thermalized, having an energy of 0.025 eV. At this point, the dominant processes are ion-molecule exothermic reactions involving ion-atom interchange or charge transfer, and ion-ion or ion-electron recombination. Clustering with polar molecules, such as water, is also likely to occur.

The ^{218}Po ions may become neutralized by removing electrons from colliding neutral molecules. Since the charge transfer process is only possible for exothermic reactions in the thermal energy range, the ionization potential of the ^{218}Po atom (8.43 eV) must be higher than that of the colliding gas molecules for the charge transfer process to take place. Busigin et al. concluded that this charge transfer process is unlikely since the ionization potential of a ^{218}Po atom is generally lower than that various gas molecules.[6] They suggested that ^{218}Po ions would be unstable and would react chemically with oxygen in air to form polonium oxide ions:

$$Po^{+} + O_2 \rightarrow PoO_2^{+}$$

Busigin et al. also estimated the ionization potential of $^{218}PoO_2^{+}$ to be about 10 eV.[6] Recently, Chu and Hopke determined the ionization potential of polonium dioxide to be 10.44 ± 0.05 eV.[7] Thus, polonium dioxide ions may be neutralized by removing electrons from trace gases having lower ionization potentials. Other chemical reactions of polonium ion with atmospheric constituents are possible.

The polonium ion may also form hydrated polonium species in the presence of water vapor.

In the present work, it is shown that the ionization potential of hydrated polonium hydroxide species is high enough to result in neutralization by removal of electrons from colliding trace gas molecules.

Wellisch found that 88% of ^{218}Po formed singly charged positive ions in a dry air system.[8] Dua et al. reported that the fraction of ^{218}Po born charged decreased with increasing relative humidity and with increasing electric field strength.[9] Busigin et al. suggested that the decrease in the fraction born charged with increasing relative humidity was due to neutralization at or just before the end of the recoil path of the ^{218}Po ions.[6] Leung and Phillips found the fraction born charged in argon at a gas age of 0.1 s to be 0.62.[10] After the ^{218}Po ions are formed, any neutralization processes which occur will affect deposition and transport of the progeny.

Porstendorfer suggested that the diffusion coefficient is a good indicator of the degree of neutralization of radon progeny.[11] The diffusion coefficient of ^{218}Po has been measured under various conditions. Busigin et al. used an electrostatic collection apparatus and found that the diffusion coefficient of neutral polonium-218 species, D_A, in argon gas increased with relative humidity.[6] At relative humidity (R.H.) = 8%, $D_A = 0.028$ cm²/s; at R.H. = 15%, $D_A = 0.043$ cm²/s, and at R.H. = 100%, $D_A = 0.046$ cm²/s. Leung and Phillips used the same method and found that the average diffusion coefficient of ^{218}Po in pure argon gas and argon containing nitrogen dioxide was about 0.03 cm²/s.

Raabe used a diffusion tube technique to determine the diffusion coefficient of ^{218}Po at various levels of R.H. Neutral ^{218}Po was found to have a diffusion coefficient of 0.047 cm²/s at R.H. = 15% and 0.034 cm²/s at R.H. = 35%.[12] Thus, the diffusion coefficient was found to decrease with R.H. Using the two filter method, Thomas and LeClare found that the diffusion coefficient increased from 0.053 cm²/s at R.H. = 0% to 0.085 cm²/s at R.H. = 20%.[13] The difference in the results of Raabe and Thomas is probably due to the difference in the age of the polonium species in their respective experiments. The age of Raabe's polonium-218 was of the order of a few minutes, whereas the average age of Thomas and LeClare's ^{218}Po was 15 to 29 s. The lower diffusion coefficients found by Raabe at higher humidities are probably the result of growth of a cluster.

Goldstein and Hopke determined the diffusion coefficient of unattached ^{218}Po in dry oxygen and in nitrogen in the presence of the trace gases n-C_4H_{10}, C_5H_{12}, NH_3, NO_2, and NO.[14] The diffusion coefficient for neutral ^{218}Po was determined to be 0.072 cm²/s, and for a mixture of charged and neutral species, 0.037 cm²/s. Frey et al. used a diffusion tube method to study the diffusion coefficient of ^{218}Po in various gas systems.[15] They found the diffusion coefficient of neutral ^{218}Po to be 0.044 cm²/s in dry N_2, 0.079 cm²/s in moist N_2 (R.H. = 80%), and 0.072 cm²/s in dry N_2 containing 10 ppm of NO_2 gas.

According to previous investigations, therefore, the diffusion coefficient of ^{218}Po changes with the experimental conditions.

The mobility has to be determined before the neutralization rate constant can be calculated. Thomas and LeClare used the Einstein equation[16] to determine the mobility from the diffusion coefficient and concluded that ^{218}Po had an ion mobility of 3.2 $cm^2/s/V$ (this corresponds to a diffusion coefficient of 0.08 cm^2/s).[13] Busigin et al. estimated the mobility of ^{218}Po ion in nitrogen or air to be 2.04 $cm^2/s/V$.[6] Using the time-of-flight method, Leung and Phillips et al. determined independent mobility data.[10] The mobility of ^{218}Po ion was found to be between 0.7 and 1.4 $cm^2/s/V$ in pure argon gas or in argon containing traces of NO_2 (<4.5 ppm).

Little work has been carried out on the neutralization rate of ^{218}Po. Busigin et al. found small ion recombination to be an important neutralization mechanism for ^{218}Po ions.[6] For neutralization of ^{218}Po ions primarily through recombination with negative small ions, Busigin et al. predicted that the recombination rate should be proportional to the square root of radon concentration.[6] This relationship was subsequently confirmed by Chu and Hopke, using a pure nitrogen system.[7]

Leung and Phillips found that the neutralization rate constant in a pure argon atmosphere was relatively independent of radon concentration at low radon concentrations (<10^5 atoms/cm^3), but proportional to the 0.6 power of radon concentration at high radon concentrations (>10^5 atoms/cm^3).[10] In an argon-nitrogen dioxide gas system, Leung and Phillips found that at radon concentrations less than 10^5 atoms/cm^3, the neutralization rate constant was independent of radon concentration, and at radon concentrations greater than 10^5 atoms/cm^3, the rate constant was proportional to the 0.3 power of radon concentration. They suggested that the dominant neutralization process at low radon concentration (<10^5 atoms/cm^3) is charge transfer and, at high radon concentrations (>10^5 atoms/cm^3), ion recombination.

George and Breslin found that the neutralization rate constant for ^{218}Po ion increased with R.H., the effect of R.H. on neutralization rate constant being most significant for R.H. < 15%. At R.H. > 15%, the neutralization rate constant became higher but varied little with relative humidity.[17]

Frey et al. studied the effect of humidity on the neutralization of ^{218}Po using diffusion coefficient measurements.[15] From diffusion coefficients determined in N_2 (0.079 cm^2/s at R.H. = 80%, 0.052 cm^2/s at R.H. = 20%, and 0.044 cm^2/s at R.H. = 0%), they concluded that the more humid was the gas system, the more completely was the ^{218}Po neutralized. They suggested that neutralization occurred by the scavenging of electrons from the recoil path of the polonium by water molecules. Goldstein and Hopke showed that the OH radical formed from the radiolysis of water molecules was a more probable electron scavenger than the water molecule itself.[14] This argument is supported by the work of Coghlan and Scott, who demonstrated inhibition of condensation nuclei formation in the presence of OH radical scavengers such as methanol or ethanol.[18] The extent of inhibition correlates well with the rate constant for reaction between the inhibiting compound and OH radicals. Thus the formation of hydroxyl radicals is argued to play an important role in both the neutralization process for the polonium ion and in the reaction chemistry leading to the formation of radiolytic nuclei.

The objective of the present study was to determine the fundamental electrostatic and diffusion properties of unattached ^{218}Po in the presence of trace hydrocarbons (hexane, ethylene, and methane) and to investigate the mechanisms of neutralization under different conditions. Hydrocarbons were chosen for study since they are commonly present in polluted air. The mechanisms of neutralization under various ionization potentials were also investigated.

EXPERIMENTAL

An electrostatic collection apparatus and a pulse-width modulated ion mobility analyzer as used previously by Leung and Phillips were used to perform the experiments in two phases.[10] In the first phase, the electrostatic collection apparatus was used to determine the diffusion coefficient, D_A; fraction born charged, F; radon concentration, C_{Rn}; and the ratio of the neutralization rate constant to the mobility, K/B, of the unattached ^{218}Po species. In the second phase, the pulse width modulated mobility analyzer was connected to the electrostatic collection apparatus to obtain independent mobility data, from which the neutralization rate constant could be calculated.

The electrostatic collection apparatus consists of two parallel aluminum discs separated by a Teflon ring. The potential of the top plate is held at ground and that of the bottom plate is varied (positive or negative) to allow collection of positively charged species on either the top or bottom plate. Neutral species are collected by diffusion. A ruggedized silicon surface barrier detector (E.G. & G. Ortec R) is located in the center of the top plate to monitor the process without disturbing the uniform electric field. The volume between the plates is 2.0 l.

Radon gas was produced by emanation from radium chloride solution in a bubbler. The R.H. was measured by a YSI dewpoint hygrometer. The pressure inside the apparatus was atmospheric and was checked by a pressure gauge for each experiment.

The flux of ^{218}Po species to the detector surface was measured by an alpha spectrometer system (Canberra). The 6.0 MeV alpha energy of ^{218}Po was separated from the 7.69 MeV alpha energy of ^{214}Po and the 5.48 MeV alpha energy of ^{222}Rn.

The electrostatic collection model described by Leung and Phillips was employed to evaluate the four parameters, D_A, F, C_{Rn}, and K/B.[10] Major assumptions of the model are: (1) the electrostatic collection apparatus has an infinitely extended parallel plate geometry since the plate diameter is much greater than the spacing, (2) the ^{218}Po ion is a single positive charge, and (3) diffusive transport and radioactive decay of the charged ^{218}Po ions under the electric field are negligible in comparison to the neutralization and electrostatic processes. Based on differential equations describing the steady state transport of charged ^{218}Po in the presence of an electric field, and steady state diffusive transport of both charged and neutral ^{218}Po species in the absence of an electric field, flux equations were obtained at different electric field strengths. The count rate, which is proportional to the flux of ^{218}Po at steady state, was fitted as a function of the applied electric field to yield the four parameters, D_A, F, C_{Rn}, and K/B. An optimization procedure based on an iterative, weighted least squares fitting routine in four dimensions was used for parameter evaluation.[19] The standard deviation for each parameter was estimated by the method given by Kalogerakis and Luus.[20] The equations involved have been described elsewhere.[10]

The pulse width modulated mobility analyzer was connected to the electrostatic collection apparatus to obtain independent mobility data. Operation of the pulse width modulated ion mobility analyzer was based on the time-of-flight method.

RESULTS AND DISCUSSION

Experiments were carried out to determine the electrostatic and diffusive properties of unattached ^{218}Po in the following gas systems: (1) trace amounts of hexane in nitrogen, (2) trace amounts of ethylene in argon, and (3) trace amounts of methane in nitrogen.

In the first phase, electrostatic collections were made, and the count rates were fitted as a function of the applied electric field to yield the four parameters: D_A, F, C_{Rn}, and K/B. The optimization procedure described previously was used. For the second phase, independent mobility data were determined by plotting the normalized count rates against the collection time, t_c.

Diffusion Coefficient of Neutral Unattached ^{218}Po, D_A

The average diffusion coefficient results for each gas system are shown in Table 1. For nitrogen as the inert gas, the average D_A was found to decrease as the ionization potential of the trace gas increased. As the ionization potential of the trace gas increases, fewer electrons will be produced along the polonium recoil path, and a lower neutralization rate and diffusion coefficient would be expected.

In gas systems of R.H. = 96 to 100%, the average diffusion coefficient was found to be 0.027 cm^2/s. This lower value is expected because the formation of clusters is significant at a high R.H.

It is shown in Figure 1 that the diffusion coefficient decreases with increasing R.H. D_A was found to have a value of 0.04 cm^2/s at R.H. = 50%.

The results for D_A are comparable to values reported by Goldstein and Hopke (1985) for other hydrocarbons, namely 0.0355 in 1.5 ppm C_5H_{10} in N_2 and 0.0365 in 16 ppb n-C_5H_{12} in N_2.

Charged Fraction

The fraction of ^{218}Po having a positive charge at the end of the recoil path, F, was found to have an average value of 0.61, in agreement with the charged fraction value reported by Leung and Phillips in an argon gas system at a gas age of 0.1 s (Table 2).[10]

Table 1. **Average Diffusion Coefficient of the ^{218}Po in Hexane-Nitrogen, Ethylene-Argon, and Methane-Nitrogen Gas Systems**

Gas System	R.H. %	I.P. (eV)	D_A(cm²/s)
13.5 ppm hexane-N2	12–17	10.18	0.047 ± 0.021
	37–43	10.18	0.043 ± -0.008
	96–100	10.18	0.029 ± 0.011
11 ppm ethylene-A	12–17	10.5	0.053 ± 0.031
	37–43	10.5	0.05 ± 0.021
	96–100	10.5	0.028 ± 0.011
10 ppm methane-N_2	12–17	12.6	0.044 ± 0.019
	37–43		0.042 ± 0.007
	96–100	12.6	0.026 ± 0.013

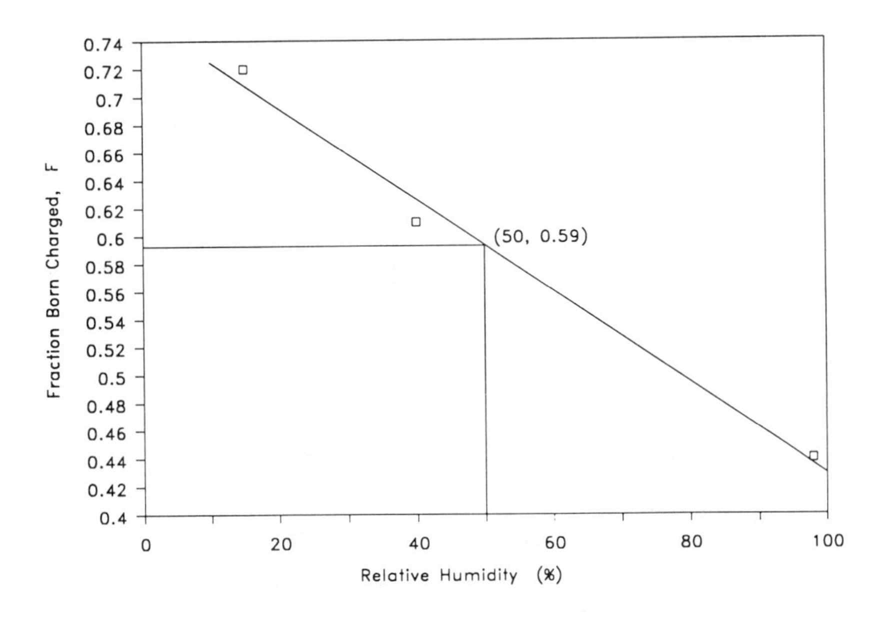

Figure 1. The effect of R.H. on diffusivity, D_A of ^{218}Po.

The positive fraction was found to decrease as the relative humidity increased (Figure 2). Since the collection time in the experiments was of the order 0.1 s, neutralization would occur between the end of recoil and the end of the 0.1 s interval. The neutralization rate was found to increase with R.H. The value of F at R.H. = 50% was determined to be 0.59 (Figure 2). The value is smaller than that reported by Porstendorfer and Mercer (0.85 to 0.88) because the collection time in the present experiments was about 0.1 s, whereas the collection time in the work of Porstendorfer and Mercer was of the order of a few milliseconds.[11] The value of F found in the present system suggests that neutralization reactions occur between the end of the recoil path and the end of the collection time interval. The age of the gas system is therefore important in determining the charged fraction.

Table 2. Effect of Relative Humidity on Charged Fraction of ^{218}Po

Relative Humidity R.H. (%)	Charged Fraction F
98	0.44
40	0.61
15	0.72

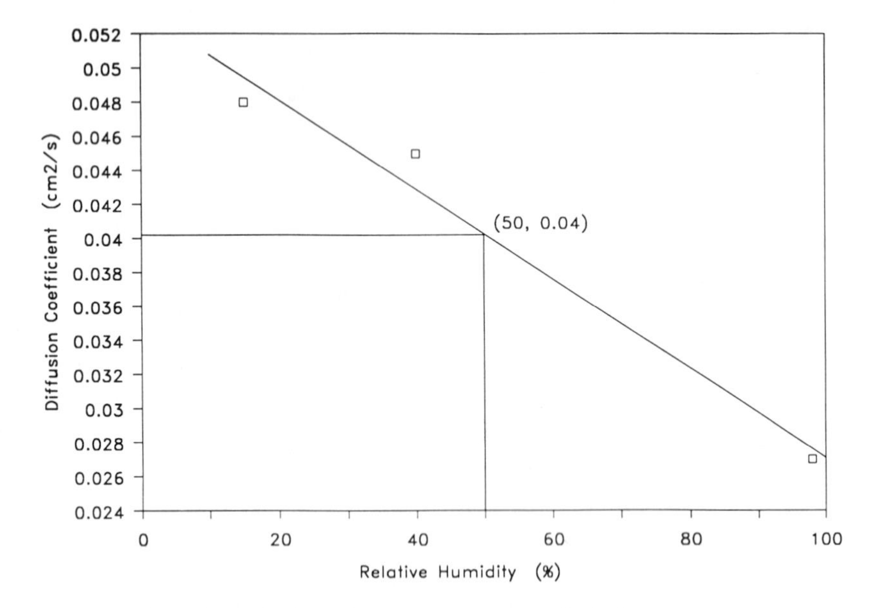

Figure 2. The effect of R.H. on fraction born charged, F of ^{218}Po.

Table 3. Effect of Relative Humidity on Mobility of ^{218}Po

Relative Humidity R.H. (%)	Mobility ($cm^2/s/V$) B
98	1.19
40	1.51
15	1.68

The average value found, 0.61, reflects the unattached charged fraction of ^{218}Po at a gas age of about 0.1 s.

Mobility of the Charged Unattached ^{218}Po, B

In the present work, the time-of-flight method was employed to determine independent mobilities. The neutralization rate was calculated for each experi-

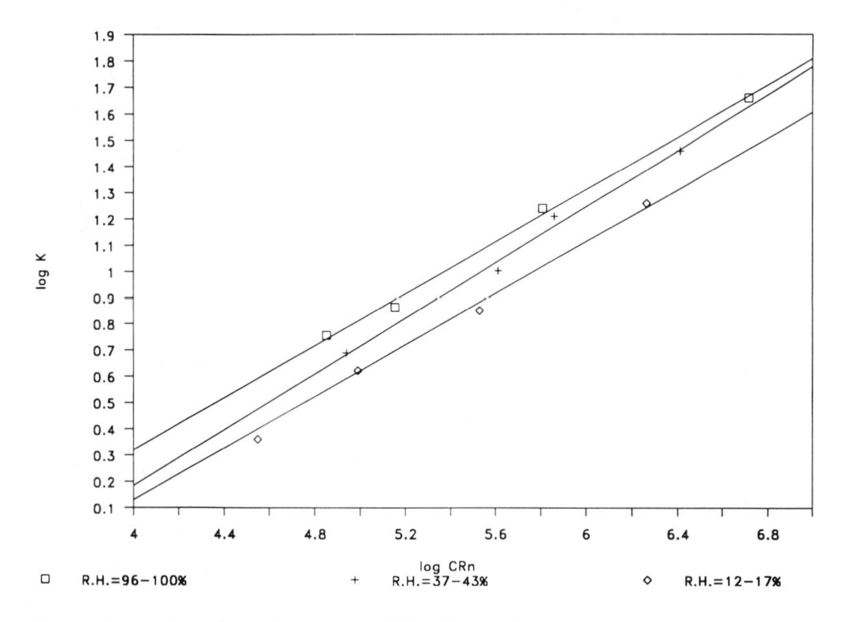

Figure 3. The effect of R.H. on mobility, B, of ^{218}Po.

ment. Mobilities of the ^{218}Po ions were determined to be 1.19 cm^2/s in a humid gas system (R.H. = 98 to 100%) and 1.68 cm^2/s/V in a relatively dry gas system (R.H. = 12 to 17%). The average value found for B was 1.51 cm^2/s/V, which is comparable to values reported by Chu and Hopke (gas system: nitrogen; 1.87 cm^2/s/V), and Leung and Phillips (gas system: argon 0.7 to 1.4 cm^2/s/V).[7,10]

Figure 3 shows that the mobility of the ^{218}Po ion decreases with increasing R.H., consistent with the occurrence of clustering, as previously reported by Leung and Phillips.[10] The mobility was found to be 1.47 cm^2/s/V at R.H. = 50%. The difference between this value and the theoretical mobility for freshly formed ^{218}Po ion (2.17 cm^2/s/V, as calculated from a mass proportionality relation), suggests that clustering and charge exchange interactions occur during the collecting time interval, which is of the order of 0.1 s.[6]

The mobility can also be calculated from the Einstein equation:[16]

$$\frac{B}{D} = \frac{e}{kT}$$

where

B = mobility
D = diffusion coefficient
e = elementary charge (= 1.60×10^{-19} coulomb)
k = Boltzmann's constant (= 1.38×10^{-23} J/K)
T = absolute temperature

Table 4. Effect of Radon Concentration and Relative Humidity on the Neutralization Rate Constant of ^{218}Po in a 10 ppm Methane (Balance Nitrogen) System

Radon Concentration (atom/cm³) C_{Rn}	Relative Humidity (%) R.H.	Neutralization Rate Constant (s^{-1}) K
5.19×10^6	100	45.7
6.42×10^5	100	17.4
1.43×10^5	97	7.3
7.09×10^4	98	5.7
2.57×10^6	40	28.8
7.24×10^6	37	16.2
4.09×10^5	42	10.1
8.68×10^4	39	4.9
1.83×10^6	13	18.2
3.38×10^5	16	7.1
9.76×10^4	15	4.2
3.54×10^4	13	2.8

Based on the diffusivity determined in the present study, 0.04 cm²/s at R.H. = 50%, the mobility estimated from the Einstein equation is 1.58 cm²/s/V, on the assumption that the ^{218}Po ion has one positive charge. The deviation between the calculated value (1.58) and the experimental result (1.47) is only 7%. An alternative interpretation is that the ^{218}Po ions are singly charged, in agreement with the literature.[11]

Neutralization Rate Constant, K, for Charged Unattached ^{218}Po Ions

Small Ion Recombination

The small ion recombination of ^{218}Po ions was studied in a methane-nitrogen gas system (Table 4). Figure 4 shows a plot of the neutralization rate constant against the radon concentration as a function of R.H. in the methane-nitrogen gas system. The correlation between the neutralization rate constant and the radon concentration at various relative humidities was found to be as follows:

$$K = 0.0217\, C_{Rn}^{0.50} \text{ at R.H. 96 to 100\%}$$
$$K = 0.0115\, C_{Rn}^{0.53} \text{ at R.H. 37 to 43\%}$$
$$K = 0.0107\, C_{Rn}^{0.52} \text{ at R.H. 12 to 17\%}$$

Correlation coefficients for linear regression were 0.998, 0.998, and 0.991, respectively. For this mechanism, Busigin et al. predicted that the neutralization rate constant should be proportional to the square root of the radon concentration, that is, $K = A\, C_{Rn}^{0.5}$.[6] If it is assumed that the neutralization of ^{218}Po ions occurs primarily through recombination with negative small ions, then the neutralization rate constant should be proportional to the negative small ion concentration. The

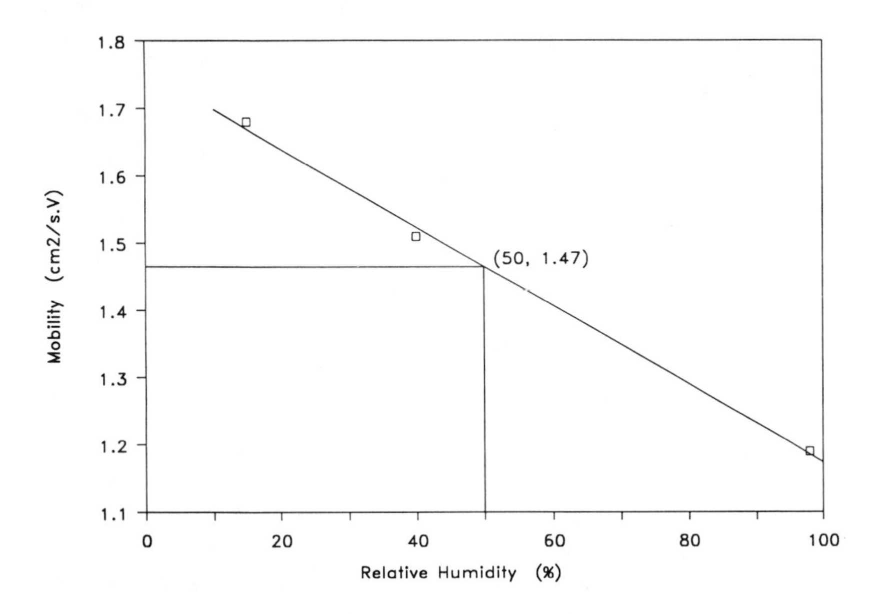

Figure 4. Log K vs. log C_{Rn}, for the methane-N_2 system (K is the neutralization rate constant and C_{Rn} is the radon concentration).

negative small ion concentration depends on the radon concentration as follows:

$$\frac{dC_-}{dt} = qC_{Rn} - rC_+C_-$$

where

q = rate constant for small ion production
r = recombination coefficient for negative and positive small ions
C_+ = concentration of positive small ions
C_- = concentration of negative small ions

If it is assumed that $C_+ = C_-$ and that steady state exists, then

$$C_- = C_+ = \left(qC_{Rn}/r\right)^{0.5}$$

Thus, the neutralization rate constant of ^{218}Po ions, K, should be proportional to negative small ion concentration, C_-, and hence to the square root of radon concentration, that is,

$$K = A\,C_{Rn}^{\ 0.5}$$

where A is a constant.

In the methane-nitrogen gas system, the 0.5 power dependence of K on C_{Rn} suggests that the neutralization of ^{218}Po ions occurs primarily through recombination with negative small ions.

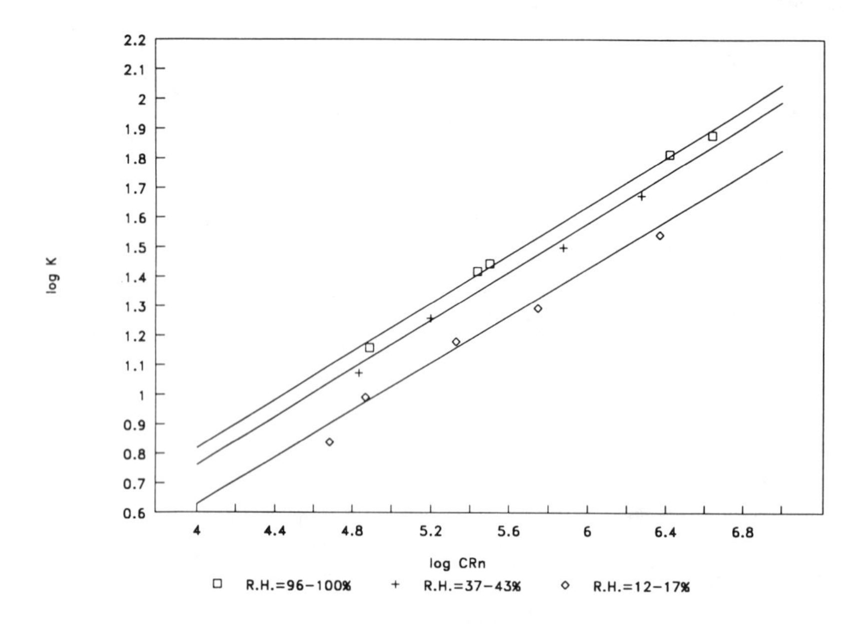

Figure 5. Log K vs. log C_{Rn}, for the hexane-N_2 system (K is the neutralization rate constant and C_{Rn} is the radon concentration).

Charge Transfer

In the hexane-nitrogen and ethylene-argon gas systems (Tables 5 and 6), the neutralization rate constant was correlated with radon concentration as follows:

hexane-nigrogen (Figure 5)

$K = 0.151\ C_{Rn}^{0.41}$ at R.H. = 96 to 100%
$K = 0.132\ C_{Rn}^{0.41}$ at R.H. = 37 to 43%
$K = 0.107\ C_{Rn}^{0.40}$ at R.H. = 12 to 17%

Correlation coefficients for linear regression were 0.998, 0.998, and 0.991, respectively.

ethylene-argon (Figure 6)

$K = 0.0497\ C_{Rn}^{0.45}$ at R.H. = 96 to 100%
$K = 0.0587\ C_{Rn}^{0.42}$ at R.H. = 37 to 43%
$K = 0.0423\ C_{Rn}^{0.43}$ at R.H. = 12 to 17%

Correlation coefficients for linear regression were 0.994, 0.994, and 0.990, respectively. The neutralization rate constant is therefore proportional to the 0.4 and 0.45 power of the radon concentration in the hexane-nitrogen and ethylene-argon gas systems, respectively. These results suggest that other neutralization mechanisms

Table 5. Effect of Radon Concentration and Relative Humidity on the Neutralization Rate Constant of ^{218}Po in a 13.5 ppm Hexane (Balance Nitrogen) System

Radon Concentration (atom/cm^3) C_{Rn}	Relative Humidity (%) R.H.	Neutralization Rate Constant (s^{-1}) K
4.33×10^6	98	75.6
2.62×10^5	100	65.2
3.15×10^5	99	27.8
2.73×10^5	100	26.2
7.69×10^4	98	14.4
1.87×10^6	43	47.2
7.48×10^5	40	31.5
1.57×10^5	42	18.1
6.81×10^4	37	11.8
2.33×10^6	16	34.8
5.52×10^5	12	19.7
2.12×10^5	17	15.1
7.36×10^4	13	9.8
4.87×10^4	13	6.9

Table 6. Effect of Radon Concentration and Relative Humidity on the Neutralization Rate Constant of ^{218}Po in a 11 ppm Ethylene (Balance Nitrogen) System

Radon Concentration (atom/cm^3) C_{Rn}	Relative Humidity (%) R.H.	Neutralization Rate Constant (s^{-1}) K
5.26×10^6	97	56.1
7.88×10^5	99	21.8
4.31×10^5	100	19.2
1.25×10^5	100	9.1
6.67×10^4	98	8.2
2.03×10^6	39	31.8
6.25×10^5	42	16.9
1.07×10^5	43	7.6
7.66×10^4	42	7.4
2.94×10^4	37	5.2
3.83×10^6	15	31.6
5.07×10^5	17	10.9
3.26×10^5	14	9.2
1.74×10^5	12	7.1
3.59×10^4	14	4.3

exist to lower the power dependence from 0.5 (for a strict small ion recombination system) to 0.4 and 0.45.

Explanation of the above effects is proposed as follows. Polonium complexes with water to form the species $Po(OH_2)_n^+$ (e.g., $Po(OH_2)_4^+$, which may reasonably

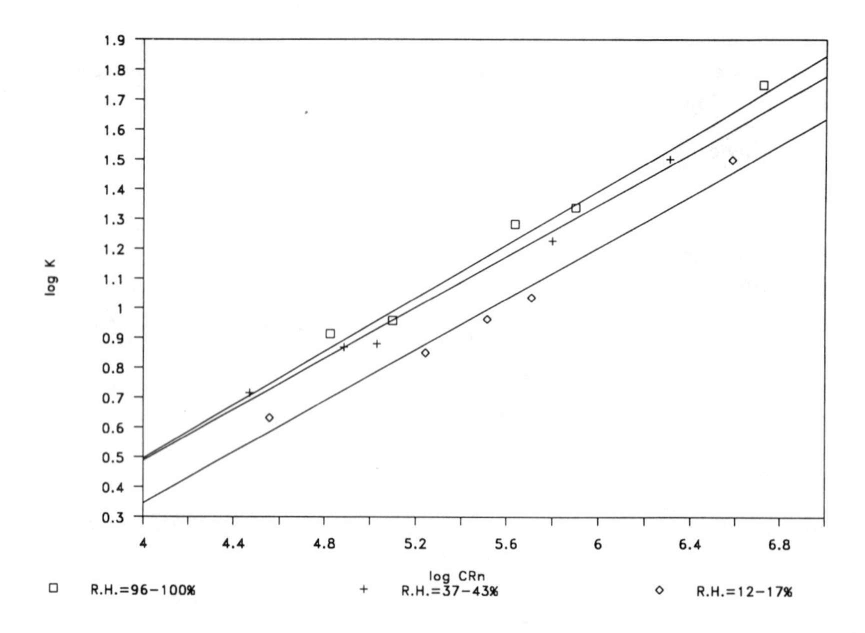

Figure 6. Log K vs. log C_{Rn}, for the ethylene-Ar system (K is the neutralization rate constant and C_{Rn} is the radon concentration).

be assumed to have a higher ionization potential than Po^+. Complexed Po ions may thus become neutralized by removal of electrons from a trace gas species of lower ionization potential, such as hexane or ethylene. According to this argument, the neutralization rate constant for charge transfer should increase with water vapor concentration. Since the ionization potential of hexane (10.18 eV; CRC Press, 1985) is lower than that of ethylene (10.5 eV), hexane is more effective in the process of charge neutralization. The ionization potential of methane (12.6 eV) is too high to allow it to be involved in charge transfer.

The power dependence of K on C_{Rn} decreased slightly below 0.5 in gas systems containing trace amounts of hexane or ethylene, but remained at 0.5 in the methane-nitrogen gas system. The decrease may be due to charge transfer involving the trace gas and the hydrated polonium species. A charge transfer reaction is possible only if the ionization potential of the hydrated polonium species is higher than that of the trace gas. The higher is the ionization potential of the trace gas, the less is the degree of charge transfer, and hence the closer is the power dependence to 0.5 (Table 7). The power dependence increases with the ionization potential of the trace gas, and levels off at 0.5 beyond a certain ionization potential, which should be the ionization potential of the hydrated polonium species. A plot of the power dependence vs. ionization potential (Figure 7) suggests that the ionization potential of the hydrated polonium species is about 10.6 eV.

Electron Scavenging

Trace gases result in electron scavenging as well as charge transfer. Hexane is a probable electron scavenger for removing electrons from the polonium recoil

Table 7. Estimate of the Ionization Potential of Hydrated Po Species: Power Exponent vs. Ionization Potential

Ionization Potential (eV)	Power Exponent
9.73	0.2
10.18	0.4
10.5	0.45
12.6	0.45

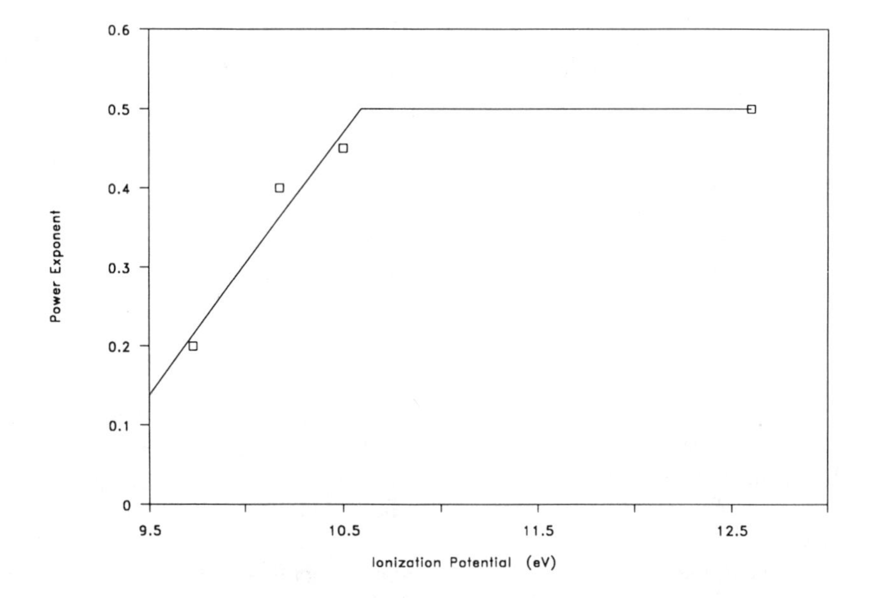

Figure 7. The estimation of ionization potential of hydrated polonium species; power exponent vs. ionization potential.

Table 8. Effect of Trace Gas Concentration on the Neutralization Rate Constant of ^{218}Po in the Hexane-Nitrogen Gas System (C_{Rn} ~ 1×10^5 atoms cm^{-3}, R.H. ~ 15%)

Hexane Concentration C_{gas} (ppm)	Neutralization Rate Constant K (s^{-1})
13.5	14.9
10	14.8
9	15.4
7.5	13.6
6	14.5
5	12.1
3	10.2
1	6.2

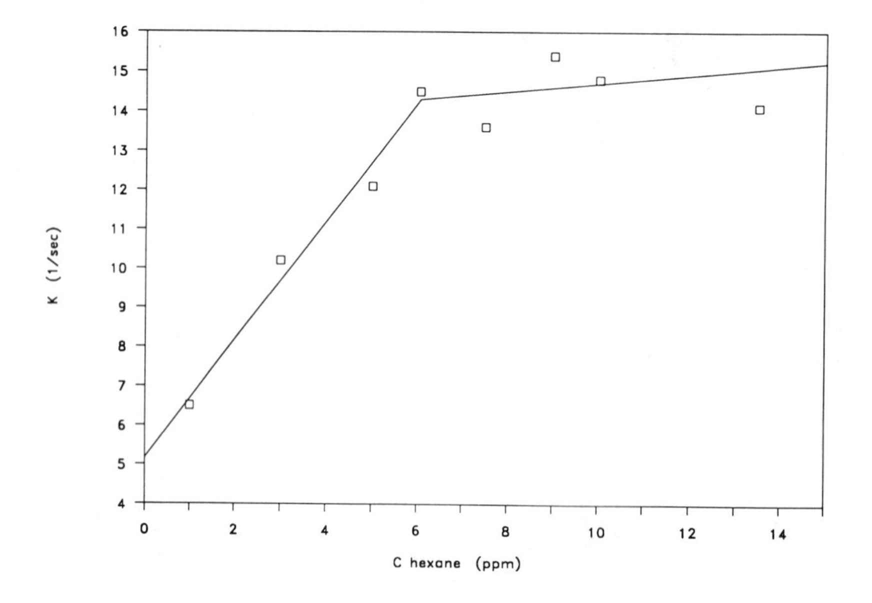

Figure 8. The neutralization rate constant, K, vs. the concentration of hexane, C_{hexane}, in nitrogen, $C_{Rn} \sim 10^{-5}$ atoms cm^{-3}, R.H. ~15%.

path and neutralizing the positive polonium ions (Table 8). A plot of the neutralization rate constant of ^{218}Po ions as a function of the concentration of hexane shows that the neutralization rate constant increases with the concentration of the electron scavenger, hexane (Figure 8). Complete neutralization is observed when the concentration of hexane in N_2 is greater than 6 ppm.

The neutralization rate constant was found to increase with relative humidity in all three gas systems, in agreement with the work of Leung and Phillips and Goldstein and Hopke.[10,14] The results suggest that water vapor is a good electron scavenger and that the electron scavenging process involving water vapor may be expressed as follows:

$$H_2O + e^- \rightarrow H + OH^-$$

$$Po^+ + OH^- \rightarrow Po + OH$$

The low electron affinity of the hydroxyl radical (1.83 eV) causes the hydroxide ion produced from water vapor radiolysis to be an excellent electron donor and a significant reactant in neutralizing polonium ions.

CONCLUSIONS

The electrostatic and diffusive properties of unattached ^{218}Po in a gas system containing trace amounts of hydrocarbons (hexane, ethylene, and methane) were determined. The diffusion coefficient has an average value of 0.04 cm^2/s, comparable to values reported by Goldstein and Hopke for other hydrocarbon systems.[14] At high R.H., the formation of clusters lowers the diffusion coefficient. In nitrogen, the diffusion coefficient decreases as the ionization potential of the trace gas increases.

The fraction of ^{218}Po having a positive charge at a gas age of 0.1 s was found to be about 0.61, in agreement with the value reported by Leung and Phillips.[10] (

The mobilities of ^{218}Po ions were determined to be 1.01 $cm^2/s/V$ at R.H. = 100% and 1.86 $cm^2/s/V$ at R.H. = 15%. The average mobility found, 1.51 $cm^2/s/V$, is in agreement with values in the literature (Busigin et al. 1981; Leung and Phillips 1987; Chu and Hopke 1988). Based on the assumption of a single positive charge on the polonium-218 ions, the mobility estimated from the Einstein equation is 1.58 $cm^2/s/V$, which is within 5% of the experimental value of 1.51 $cm^2/s/V$. This result may alternatively be interpreted as indicating that the ^{218}Po ions are singly charged.

The neutralization of unattached ^{218}Po positive ions is a complicated process and is controlled by different mechanisms under different conditions. In all three gas systems, the neutralization rate constant was found to be proportional to an exponential power of the radon concentration, namely 0.4, 0.45, and 0.5 for hexane-N_2, ethylene-Ar, and methane-N_2, respectively. The results suggest that small ion recombination plays an important role in the neutralization of the unattached ^{218}Po positive ions. The charge transfer mechanisms involving hydrated polonium species and trace gases (hexane or ethylene) lower the power dependence of the neutralization rate constant on radon concentration to less than 0.5. Since the charge transfer neutralization rate increases with the concentration of H_2O molecules, an increase in radon concentration causes a decrease in the charge transfer neutralization rate due to the increasing rate of radiolytic decomposition of water vapor. The presence of methane as trace gas does not cause charge transfer because of the relatively high ionization potential of methane (12.6 eV). In all three gas systems, the neutralization rate constant was found to increase with relative humidity, suggesting that the electron scavenging process involving water vapor radiolysis is an important neutralization mechanism.

Hydrated polonium species lowers the power dependence of K on C_{Rn} by undergoing charge transfer with hexane and ethylene, but not with methane, whose ionization potential is too high. A plot of the power dependence vs. ionization potential leads to an estimate of 10.6 eV for the ionization potential of the hydrated polonium species.

ACKNOWLEDGMENTS

This work was supported by strategic and operating grants from the Natural Sciences and Engineering Research Council of Canada and by a contract from the Atomic Energy Control Board of Canada.

REFERENCES

1. Ham, J.M. (Commissioner), Report of the Royal Canadian Commission on the Health and Safety of Workers in Mines, (Ministry of the Attorney-General, Province of Ontario, Canada, 1976).
2. Nero, A.V., "Indoor Radiation Exposures from ^{222}Rn and its Daughters: A View of the Issue," *Health Phys.,* 45:217 (1983).
3. National Council on Radiation Protection: Report No. NCRP-45, (1975).
4. National Council on Radiation Protection: Report No. NCRP-77, (1984).
5. Jacobi, W., "Activity and Potential Alpha-Energy of ^{222}Radon- and ^{220}Radon-Daughters in Different Air Atmospheres," *Health Phys.,* 22:441 (1972).
6. Busigin, A., van der Vooren, A.W., Babcock, J., and Phillips, C.R., "The Nature of Unattached RaA (^{218}Po) Particles," *Health Phys.,* 40:333 (1981).
7. Chu, K.D. and Hopke, P.K., "Neutralization Kinetics for Polonium-218," *Environ. Sci Technol.,* 22:711 (1988).
8. Wellisch, E.M., "The Distribution of the Active Deposit of Radium in an Electric Field — II," *Philos. Mag.,* 26:623 (1913).
9. Dua, S.K., Kotrappa, P., and Gupta, P.C., "Influence of Relative Humidity on the Charged Fraction of Decay Products of Radon and Thoron," *Health Phys.,* 45:152 (1983).
10. Leung, H.M.-Y. and Phillips, C.R., "The Electrical and Diffusive Properties of Unattached ^{218}Po in Argon Gas," *Radiat. Prot. Dosimetry,* 18:3 (1987).
11. Porstendorfer, J. and Mercer, T.T., "Influence of Electric Charge and Humidity upon the Diffusion Coefficient of Radon Decay Products," *Health Phys.,* 37:191 (1979).
12. Raabe, O., "Measurement of the Diffusion Coefficient of RaA," *Nature (London),* 217:1143 (1968).
13. Thomas, J.W. and LeClare, P.C., "A Study of the 2-Filter Method for Radon-222," *Health Phys.,* 18:113 (1970).
14. Goldstein, S.D. and Hopke, P.K., "Environmental Neutralization of Polonium-218," *Environ. Sci. Tech.,* 19:146 (1985).
15. Frey, G., Hopke, P.K., and Stukel, J.J., "Effect of Trace Gases and Water Vapor on the Diffusion Coefficient of Po-218," *Science,* 211:480 (1981).
16. Fuchs, N.A., *The Mechanics of Aerosols,* (Oxford, England: Pergamon Press Limited, 1964).
17. George, A.C. and Breslin, A.J., "Measurements of Environmental Radon with Integrating Instruments," (Presented at Atomic Industrial Forum Uranium Mill Monitoring Workshop, Albuquerque, NM, 1977).
18. Coghlan, M. and Scott, J.A., "A Study of the Formation of Radiolytic Condensation Nuclei," in *Prod. in Atmospheric Electricity,* Ruhnke, L.H., and Latham, J., (Eds.), (Hampton, VA: A. Deepak Publishing, 1983).
19. Luus, R. and Jaakola, T.H.I., "Optimization by Direct Search and Systematic Reduction of the Size of Search Region," *A. I. Ch. E. J.,* 19:760 (1973).
20. Kalogerakis, N. and Luus, R., "Sequential Experimental Design of Dynamic Systems Through the Use of Information Index," *Can. J. Chem. Eng.,* 62:730 (1984). *CRC Handbook of Chemistry and Physics,* 66th ed., Weast, R.C. and Astle, M.J., (Eds.), (Boca Raton, FL: CRC Press, 1985).
21. Busigin, C., Busigin, A., and Phillips, C.R., "The Chemical Fate of ^{218}Po in Air," in *Proc. Int. Conf. on Radiation Hazards in Mining,* Gomez, M. (Ed.), (Golden, CO: Soc. of Mining Engineers, Inc.), p. 1043.

CHAPTER 14

Sorption of Radon on Porous Materials and the Importance of Controlling Radon in the Indoor Environment

Stephen D. Schery and Tracie L. Lopez

ABSTRACT

Sorption coefficients have been measured for a range of porous materials common in the indoor environment. Coefficients can vary widely, by factors of a hundred or more, but generally are small compared with well-known adsorbents such as activated charcoal. Almost all coefficients decrease with increasing temperature and moisture. An important consequence of the presence of sorption appears to be its role in retarding the transport of radon through porous materials. Since sorption can also cause a strong temperature and moisture dependence in the effective diffusion coefficient, it should be more often explicitly incorporated in radon transport models than is presently the case. Other identified, but less likely, consequences of sorption include its ability to cause pulsed releases of radon to indoor air, and to act as a reservoir for radon reducing the concentration of radon in the indoor air.

INTRODUCTION

The subject of sorption of gases on porous materials is proving important in a number of environmental and pollution contexts. Sorption can be important for retarding transport of pollutant vapors through soil[1,2] as well as release or retention of pollutants in indoor air (see other chapters of this volume). Despite a long history of radon gas as a pollutant, first in mine atmospheres and more recently in indoor air, surprisingly little study has taken place on the sorption properties of radon on porous materials common in the indoor and outdoor environments. In

1980 Tanner[3] reviewed the limited data available on radon sorption and recommended that this was an important subject requiring greater priority in radon research. Several more recent papers have presented evidence for desorption as an important mechanism enhancing release of radon from certain materials (concrete, soil, tailings) with increasing temperature and moisture.[4,5] Unfortunately, the experimental procedures used in those papers did not directly measure sorption, so it was difficult to evaluate and quantify the specific role of sorption in the reported observations. Responding to this need for sorption specific measurements, Schery and Whittlestone[6] developed a procedure for directly measuring sorption of radon on porous materials and reported results for soils and rocks common in the outdoor environment.[7] Certain situations were identified where sorption might play an important role in release of radon to the atmosphere. These situations included unvegetated soils in arid and semi-arid climates and dry, loose materials (rocks, etc.) at the extreme surface of the earth. However, it was pointed out that the moisture content of most soils is significant, particularly below the surface skin. Based on the measured moisture dependence of sorption coefficients, this moisture content of soils would reduce the importance of sorption over much of the surface of the earth.

Since the indoor environment would normally be expected to be drier than the outdoor environment, the indoor environment seems a particularly suitable place to look for a larger role of sorption in the transport of radon through porous materials. Furthermore, the indoor environment is of particular interest in light of the recent awareness of radon as an important indoor pollutant. The goal of the research reported in this paper is to extend the measurements of Schery and Whittlestone[7] to a representative cross section of porous materials important for radon transport in the indoor environment, and see if there are any circumstances under which sorption is important for control of radon in indoor air.

EXPERIMENTAL PROCEDURE

The experimental technique is that developed by Schery and Whittlestone.[6,7] Air containing trace amounts of radon is passed over milliliter-size samples of porous media until equilibration is reached between radon in the air of the pore space and radon sorbed on the grains of the porous media. Samples are counted in a gamma detector, and from the count rate from radon and its progeny, and from the concentration of radon in the pore air, the sorption coefficients can be deduced. Calculation of the sorption coefficients requires a correction for the radon residing in the pore space of the samples based on separate measurements of porosity or solid grain volume.

For the present measurements, radon-enriched air was produced by passage of dried air through a commercial 290-kBq ^{226}Ra source (Pylon Ltd., Canada). The air was filtered, then passed through a series of 7-ml vials that were in a temperature-controlled chamber. The vials were made of glass with a brass top sealed by a small O-ring in order to minimize sorption in the containers themselves. A constant-flow pump at the entrance to the system provided flow rates in the range of 1 to 10 ml min^{-1} while maintaining an overpressure in the system to minimize inward leaks. At these low flow rates the radon activity was high enough that no correction was necessary for the natural radioactivity content of samples. Moisture content of the samples was controlled by direct addition of water or by bubbling of the radon-rich air through a vial of water placed before a sample. Samples were removed one at a time from the temperature controlled chamber, were counted for 5 to 10 min in a 5-cm diameter by 8-cm depth NaI well detector with a gate set to count all gamma rays above 85 keV, were weighed for moisture content, and then were returned to the chamber.

The sorption coefficient k (ml g^{-1}, the volume of air in contact with the sample that would contain the same activity as sorbed in a unit mass of the sample) was deduced by comparison of the counts for an unknown sample to counts from two reference samples in the same sample set. Usually the reference vials consisted of an empty vial, and a vial of smooth glass beads of known porosity and negligible sorption. For a given temperature and moisture, measurements were repeated until steady-state results were obtained. Typically, 12 to 24 h were required to reach the steady-state, presumably reflecting the time necessary for diffusion of radon through the samples, sorption of radon by the grains, and ingrowth of radon progeny which provide most of the gamma rays which were detected.

The porosity or solid grain volume of samples was determined separately using the gas expansion method.[8] Air was used as the gas with small displacements from atmospheric pressure and with equilibration times less than a minute. Moisture content was determined by comparing wet weight to dry weight obtained by drying to constant weight in an oven at 80°C. Measurements were carried out at atmospheric pressure averaging about 85 kPa at the Socorro, New Mexico laboratory. The accuracy of the technique was checked by measuring the sorption coefficient of water as a function of temperature (for this case the sorption mechanism is solution and the bulk volume constitutes the "grain" volume). The results compared with the accepted values of k for water[9] are shown in Figure 1. Within error bars, reflecting repeatability of measurements at a given temperature, good agreement is obtained. For the routine measurements reported in this paper, the absolute error is estimated to be about ±25%. Greater accuracy was possible but not felt warranted given the difficulty of characterizing the chemical composition, pore structure, etc., of complex porous environmental materials with equivalent accuracy.

Auxiliary measurements were made on some samples to provide more information on their physical and chemical structure. Internal surface area was measured at liquid nitrogen temperature with a commercial instrument (Monosorb Surface Area Analyzer, Quantachrome, Model MS-13) using 30-mol percent nitrogen in helium carrier gas.

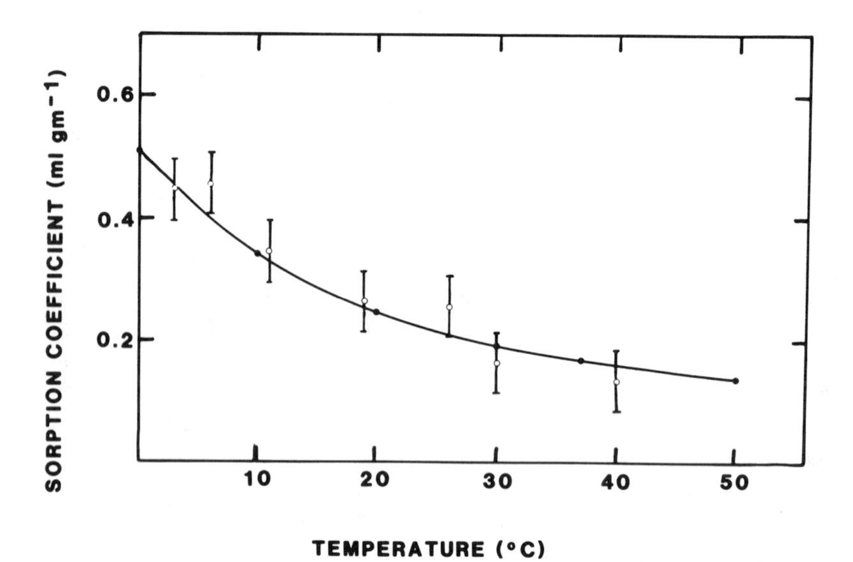

Figure 1. Measurements of the sorption coefficient for water (open circles) compared with the accepted values (filled circles) as a function of temperature. The solid curve is an approximate visual fit to the accepted values.

RESULTS

Representative results for sorption coefficients of dry samples as a function of temperature are shown in Figure 2. In most cases, the sorption coefficient decreased with increasing temperature. There are a few exceptions, such as the cases for certain wood products such as dried pine (lumber) and paper (newsprint). This anomalous behavior might be the artifact of some unknown indirect effect of temperature in our procedure (for example, the time needed to achieve the steady-state at different temperatures was checked) but was persistent over many replications.

Representative results for sorption coefficients as a function of moisture at a fixed temperature of 26°C are shown in Figure 3. A trend of decreasing sorption with increasing moisture is evident. This trend was seen with all samples over the range of moistures studied (generally up to a maximum moisture content less than 25%).

Further information is tabulated in Table 1 for those samples for which sufficient data were obtained to establish a temperature and moisture dependence. The sorption coefficient (k) data for each sample were fit with the two-dimensional, three-parameter, linear function

$$k = k_o + k_t T + k_m M \qquad (1)$$

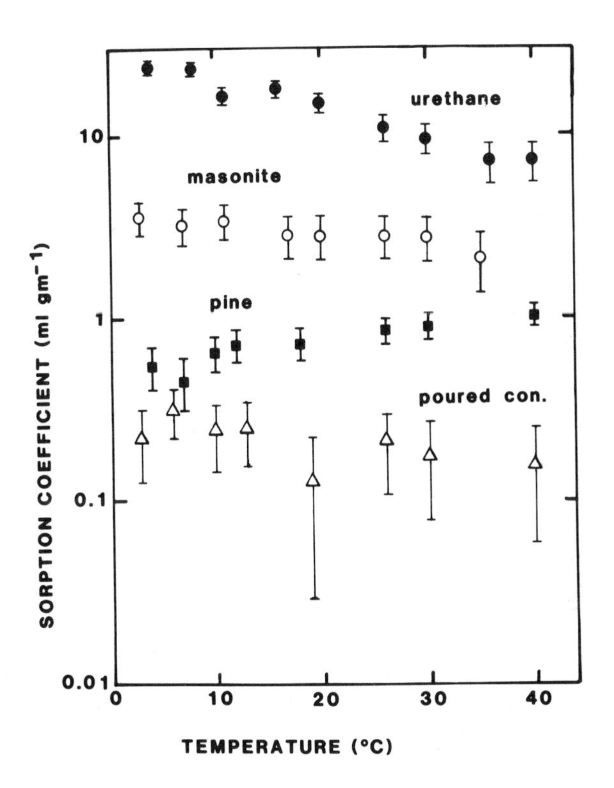

Figure 2. Sorption coefficient versus temperature for dry (<0.2%) samples of urethane foam, masonite, pine lumber, and poured (slab) concrete.

where T is temperature in degrees Celsius and M is percent moisture by weight. This function is clearly a general low-order approximation and does not represent particularly well, for example, the apparent exponential dependence in some of the data when the temperature and moisture range is wide. However, the use of more sophisticated functional forms was considered presumptive without more complete knowledge about the specific sorption mechanisms involved.

The proportions between radon sorbed on grains and radon residing in the pore space for some selected conditions of moisture and humidity are tabulated in Table 2.

This proportion, labeled r, was calculated from

$$r = \frac{k\rho}{\varepsilon} \tag{2}$$

where ρ is the density and ε the porosity of each sample. Values for k were determined by interpolation and use of Equation 1. The density or porosity of some materials (urethane foam, concrete, etc.) can be quite variable as used in the indoor environment. Due to this variability and the use of interpolation for k, the entries in Table 2, while representative, are less reliable than Table 1 and Figures 1 and 2 as primary reference data.

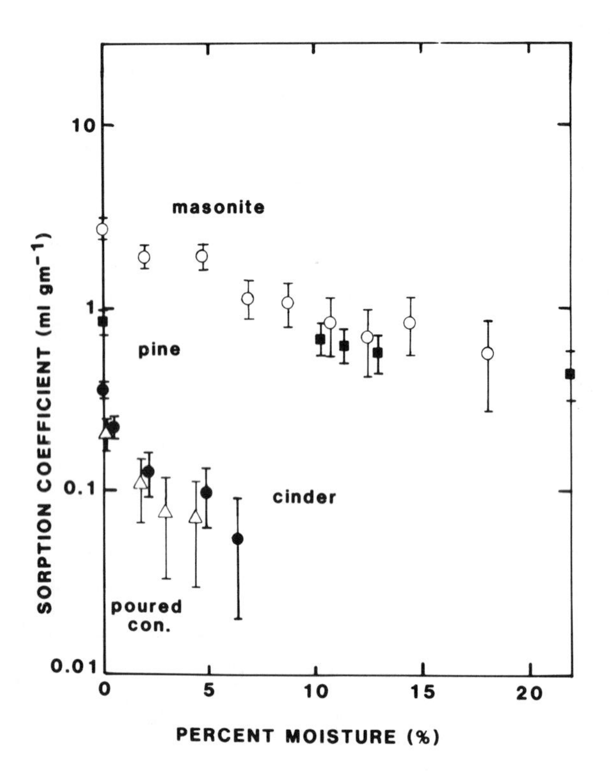

Figure 3. Sorption coefficient at a temperature of 26°C vs. moisture for samples of masonite, pine lumber, concrete cinder block, and poured concrete.

Figure 4 shows a plot of sorption coefficients for dry samples at 0°C vs. internal surface area for samples for which measurements were available. Also shown (filled circles) are data for outdoor materials from Schery and Whittlestone[7] to provide a larger sample size for comparison and help in the interpretation of the data.

ANALYSIS

A wide variability in sorption coefficients, by a factor of 100 or more is indicated in Table 1. Even so, compared with well known high adsorbents such as activated charcoal, where sorption coefficients are often in the range of many thousands of milliliters per gram at room temperature, these sorption coefficients are generally small. Hence it is not obvious how important sorption is for control of radon in indoor air. It is necessary to take a closer look at specific mechanisms by which sorption could affect radon transport to indoor air.

One simple approach is to ask to what extent sorption is a significant reservoir or sink for radon compared with the radon stored in the bulk room air of a building. For example, consider a room 5 m × 5 m × 3 m with walls, ceiling, and floor of

Table 1. Results of a Three Parameter Fit to the Sorption Data

Material	k_o ml/g	k_T ml/g·T	k_M ml/g·M	Range T (°C)	Range M (%)	Comments
Concrete cinder block	0.37	–0.0006	–0.051	4–40	0–6.5	Porosity = 25%, cinders, gypsum, and concrete
Fired brick	0.096	0.001	–0.045	0–36	0–0.5	
Pumice concrete	0.25	–0.003	–0.049	0–36	0–3	Porosity = 21%, pumice, gypsum, and concrete
Sheetrock	0.19	–0.003	–0.012	2–36	0–5.5	Gypsum with paper binding
Oven dried pine	0.49	0.014	–0.019	2–40	0-25	Lumber sample
Urethane foam	27.1	–0.54	–1.20	4–40	0–9	
Particle board	2.05	–0.010	–0.086	3–40	0–14	Wood particles in cured resin binding
Masonite	3.95	–0.052	–0.13	2–39	019	Steam exploded wood fibers
Poured concrete	0.29	–0.004	–0.034	3–40	0–5	Gravel in cement binding, porosity = 17%
Oak	2.47	0.075	–0.045	0–30	0–9	Lumber sample
Celotex	9.21	–0.002	–0.084	0–30	0–20	Wood particles with asphalt binding
Fiberglass wall insulation	1.64	–0.012	–0.044	5–36	0–16	Glass fibers, most diameters 3 to 7 μm
Nylon carpet	1.55	0.006	–0.034	5–36	0–7	Fibers about 50 μm diameter, ANSO IV brand
Paper	5.3	0.06	–0.23	7–25	0–20	Newsprint

Table 2. Values of $r = k\rho/\varepsilon$ for Different Conditions

	Dry Warm Conditions (0%, 22°C)	Dry Cool Conditions (0%, 12°C)	Equilibrated in Air at 48% RH and T = 22°C	
	r	r	% Moisture	r
Concrete cinder block	2.8	2.8	0.8	2.5
Fired brick	1.9	1.7	0	1.9
Pumice concrete	1.6	1.9	0.9	1.2
Sheetrock	0.15	0.18	2.4	0.11
Oven dried pine	0.67	0.55	11.0	0.49
Urethane foam	0.32	0.44	0	0.32
Particle board	2.8	3.0	9.1	1.6
Masonite	5.6	6.7	8.1	3.5
Poured concrete	2.7	3.2	1.4	2.1
Oak	5.7	4.6	9.8	5.1
Celotex	3.2	3.2	6.9	3.0
Fiberglass wall insulation	0.23	0.25	0	0.23
Nylon carpet	0.16	0.16	1.9	0.16
Paper	9.0	8.2	12	5.3

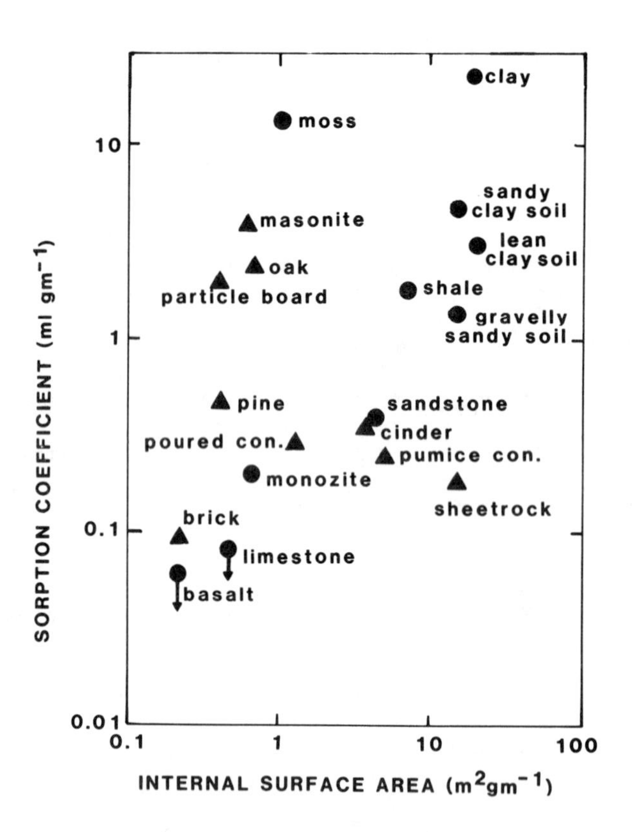

Figure 4. Internal surface area vs. sorption coefficients at 0°C and 0% moisture for samples for which surface area measurements were available. The filled circles are data from Schery and Whittlestone.[7]

concrete 0.05 m thick. Assume the concrete has a sorption coefficient of about 0.3 ml g^{-1} (a representative value, see Table 1) and a density of about 2 g ml^{-1}. The volume of the concrete is 5.5 m^3, so from the standpoint of sorption of radon it provides an equivalent reservoir volume of 5.5 $m^3 \times 2$ g $ml^{-1} \times 0.3$ ml $g^{-1} = 3.3$ m^3 compared with a bulk volume of room air of 75 m^3. Hence, assuming good radon exchange and steady-state conditions, most of the radon would reside in the room air (about 96%). The actual proportion in the concrete is likely to be even less than this estimate implies due to a reduction in the sorption coefficient because of moisture likely to be present and due to restricted penetration of radon through the concrete because of the finite diffusion length of radon. Calculations for standard sized rooms and the other materials from Table 1, for volumes of materials likely to be present in construction, generally also indicate a small role of sorption in creating a significant reservoir (e.g., masonite and urethane foam have much larger sorption coefficients but the mass of such materials used in construction would typically be much smaller, which would tend to compensate for the larger sorption coefficients). Special situations can be envisaged where sorption is more important. A small closet or cabinet filled with wood products (such as books or paper) might well have sorption the dominant reservoir of radon. Dry soils, depending on clay content, can have significant sorption coefficients.[7] Thus, thick-walled adobe houses such as those that occur in the southwest U.S. might have sorption in the mud walls a more significant reservoir of radon. Situations have to be evaluated on a case-by-case basis. However, circumstances where sorption provides the dominant reservoir seem the exception, and the more general rule seems to be that usually sorption coefficients are too small and the volume-to-surface area of rooms too large for the building materials and furnishings to be a major reservoir of radon.

Another approach to analyzing the significance of sorption is to look at transport of radon within the porous materials themselves. Some important information is provided in Table 2. Even at the reduced sorption present with typical moisture in materials, the majority of materials have more radon in the sorbed state than in the pore air itself. The consequences can be indirect but potentially significant, for this situation indicates that diffusive transport may be controlled more by the properties of sorption than strictly by the physical laws for three-dimensional diffusion through the pore air itself (as sometimes assumed). Although diffusion can still be represented by a single effective diffusion coefficient, diffusion will tend to be slowed by sorption and have an altered temperature and moisture dependence. The above arguments can be quantified by looking at the results of a simplified model for the effective diffusion coefficient (D_e) when both pore diffusion (D_p) and surface diffusion (D_s) are present.[10] In our present context the result is

$$D_e = \frac{D_p + D_s \frac{k\rho}{\varepsilon}}{1 + \frac{k\rho}{\varepsilon}} \tag{3}$$

Usually surface diffusion is much slower than pore diffusion ($D_s << D_p$) so

$$D_e = \frac{D_p}{1 + \frac{k\rho}{\varepsilon}} \tag{4}$$

is a good approximation. Equation 4 highlights the sensitivity of the diffusion coefficient to the sorption coefficient. Large sorption coefficients will significantly reduce D_e. Furthermore, the dependence of k on temperature and moisture seen in much of the present data will cause stronger dependences of D_e on these variables than would otherwise be present with pure molecular pore diffusion. A porous slab possessing significant sorption could be a barrier to radon diffusion when dry but might become permeable to radon when moist due to rapid decrease in k with increasing moisture. For example, Equation 4 and the data of Figure 3 indicate the diffusion coefficient for cinder block could be reduced by a factor of over four when slightly moist compared with when dry.

Another consequence of the dependence of the effective diffusion coefficient on the sorption coefficient is the possibility of seeing pulsed releases of radon from building materials and furnishings upon change in temperature or moisture. Proper modeling of this situation requires analysis of changes in the concentration of radon in the pore space, and time dependence of the moisture and temperature penetration, as well as variation of D_e.[7] Such detailed modeling is outside the scope of this chapter, but there are a few simpler calculations that can provide insight. Consider an increase in indoor temperature of 10°C, say from 12 to 22°C, which might be a relatively feasible indoor event. For the room with concrete walls analyzed earlier, a representative sorption coefficient might change from of the order of 0.35 to 0.30 ml g^{-1}. If this change occurred instantaneously, and the desorbed radon was uniformly distributed throughout the indoor air, the increase in room air radon would only be of the order of 1%. Likewise, Table 2 indicates that such a 10°C temperature change is only sufficient to change most r-values by a small fraction, suggesting a rather small change in radon in the pore space and in the diffusive transport to the surface. Larger temperature changes, significant changes in moisture, or release into smaller air spaces would of course be capable of producing more noticeable changes in the radon level. If the instantaneous exhalation *rate* from a surface was measured, rather than the integrated effect on room air, effects could be more noticeable. On balance, though, it does not appear major pulsed releases would be common.

Detailed analysis of the physical mechanisms involved in radon sorption is outside the scope of this chapter. However, it is interesting to note that if organic materials are excluded (moss and wood products), as shown in Figure 4, there is a reasonable correlation between sorption coefficients and internal surface area as determined by the nitrogen adsorption procedure. This result might suggest the importance of simple adsorption for these inorganic materials. The fact that the organic materials form a separate grouping might reflect a different physical mechanism for sorption, or perhaps just some measurement artifact such as that resulting from internal surface area measurement at very cold temperatures.

CONCLUSIONS

Sorption coefficients have been measured for a cross section of materials common in the indoor environment. Coefficients vary widely, by a factor of 100 or greater, but compared with well-known adsorbents, such as activated charcoal, are relatively small. Coefficients generally decrease with increasing moisture and temperature. Given the generally dryer conditions in the indoor environment than outside, there appears a greater chance for sorption to play a significant role inside than outside.

Analysis indicated that the most likely effect of sorption is to change the diffusion process through indoor materials from simple molecular diffusion through the pore space to a more complex process involving interaction between pore diffusion and sorption on grains. One likely consequence is that effective diffusion coefficients would have a stronger dependence on moisture and temperature than would otherwise be the case. Slabs that are barriers to radon at one temperature and moisture might not be barriers at other temperatures and moistures. This effect of sorption probably needs to be better taken into consideration in transport models and computer codes that are being developed.

Only in special situations did it appear sorption would be strong enough to give observable pulses of radon in the indoor air of rooms with changing temperature. Similarly, it appeared that generally sorption coefficients were too small, and the volume to surface ratio of rooms too big, for sorption to provide a major reservoir for radon compared with the bulk volume of the room air. Some exceptions were noted. Enclosures with high surface-to-volume ratios, for example closets or cabinets filled with paper products, might have sorption a significant reservoir of radon.

ACKNOWLEDGMENTS

This research was supported in part by the U.S. Department of Energy grant DE-FG04-88ER60662. The assistance of Karen West in data collection is appreciated.

REFERENCES

1. Kreamer, D.K., Weeks, E.P., and Thompson, G.N., *Water Resour. Res.,* 24:331–341 (1988).
2. Raub, J.A. and Grant, L.D., Paper 89-53.2, The 82nd Annual Meeting and Exhibition of the Air Waste Management Association, June 25-30, 1989 (Anaheim, CA, 1989).
3. Tanner, A.B., in *Natural Radiation Environment III*, Gesell, T.F. and Lowder, W.H., (Eds.), (Springfield, VA: National Technological Information Service, 1980), pp. 5–56.
4. Stranden, E., *Health Phys.,* 47:480–484 (1984).
5. Gan, T.H., Mason, G.C., Wise, K.N., Whittlestone, S., and Wyllie, H.A., *Health Phys.,* 50:407–411 (1986).
6. Schery, S.D. and Whittlestone, S., in *Geologic Causes of Natural Radionuclide Anomalies: Proceedings of the GEORAD Conference,* Marikos, M.A. and Hansman, R.H., (Eds.), (Rolla, MO: Division of Geology and Land Survey Special Publication No. 4, Missouri Department of Natural Resources, 1988), pp. 131–139.
7. Schery, S.D. and Whittlestone, S., *J. Geophys. Res.,* 94:18297–18303 (1989).
8. Collins, R.E., Chapter 1, in *Flow of Fluids Through Porous Materials,* (Tulsa, OK:

Petroleum Publishing Co., 1976).

9. "Ionizing Radiation: Sources and Biological Effects," United Nations Scientific Committee on the Effects of Atomic Radiation, New York, United Nations, E.82.IX.8, 06300P (1982).
10. Ash, R., Barrer, R.M., and Lawson, R.T., *J. Chem. Soc., Faraday Trans. I*, 69:2166–2178 (1973).

CHAPTER 15

Criteria for Closed Chamber Measurements of Radon Emanation Rate

L. Morawska and C.R. Phillips

ABSTRACT

Determination of the radon emanation rate from solid materials has been the objective of many studies. The experimental method most often used for determining the radon emanation rate is the so-called "closed chamber" method, in which the growth of radon concentration in a closed chamber is used as the basis for the calculation. Usually, only certain regions of the concentration growth curve are used for calculation purposes, specifically, the initial region of the curve, or the equilibrium value to which it converges. In spite of the experimental simplicity of this method, the mathematical procedures used to calculate the free emanation rate (defined as the emanation rate to an infinite space of air) raise many questions, the answers to which are often contradictory from one implementation of the method to another. The objective of this chapter is to provide quantitative values of the inherent bias in determining the emanation rate for various experimental conditions in the closed chamber method. Quantitative comparison is made among the mathematical models most often applied. An assessment is provided of the usefulness of each model for its assumed geometry of measurement.

INTRODUCTION

Knowledge of the radon emanation properties of a material is important for many practical purposes as well as being of fundamental scientific interest. The experimental methods commonly used to investigate radon emanation may be divided into two groups. To the first group belong methods based on the accumulation of radon emanated from the sample into an emanation chamber. This is the so-called static method. The principle of the second group is dynamic and is based on an air or gas flow over the surface of the sample to collect radon emanating therefrom.

The parameters that characterize emanation are emanation coefficient (sometimes called — with less precision — emanating power) and emanation rate. The radon emanation coefficient is defined as the ratio of radon atoms born in the material and released into the free pore space to all of the radon atoms born in the material. The emanation coefficient depends on the radium concentration and distribution in the material, the material porosity, and on what fills the inner pore volume (water or air). The emanation coefficient is not affected by external factors such as temperature, pressure, or radon concentration outside the sample.

The emanation rate characterizes the flux of radon from the material to the outer air volume. The emanation rate can be expressed as the total emanation rate of a sample [Bq s^{-1}], the mass emanation rate of a material [Bq $kg^{-1}s^{-1}$] and the surface emanation rate [Bq $m^{-2}s^{-1}$]. In the case of so-called free emanation, in which the volume of the air surrounding the sample is infinite relative to the volume of the sample, correlation between the emanation coefficient and the emanation rate depends upon the diffusion length and the dimensions of the sample. However, unlike the emanation coefficient, the emanation rate is affected by external agents such as temperature and pressure changes. If the volume of the air surrounding the sample is not infinite and the radon concentration outside the sample cannot be assumed to be zero, the radon emanation process taking place cannot be referred to as free emanation. It is described as bound emanation. The bound emanation rate is lower than the free emanation rate. From the point of view of applications, it is important to know the free emanation rate.

The experimental method most often used for determination of the emanation rate is the closed chamber method, which, in spite of its simplicity, has a substantial disadvantage, namely, that the measured emanation rate is often not the free emanation rate. The closed chamber method is based on measurement of radon concentration as a function of time in a chamber enclosing the sample. This measurement method is not new. In different forms, it has been applied for the past few decades. Krisiuk et al. present mathematical expressions for the time- and position-dependent concentration of radon in the material and provide a time-dependent solution to this problem.[1] As shown by Krisiuk et al. and Samuelsson and Pettersson, it is possible on the basis of this theory to calculate the time-dependent emanation rate and the radon concentration growth in the sample chamber.[1,2] The derived expressions are so complicated, however, that in practice it is not possible to determine the change in the emanation rate from the experimental data on radon concentration in the chamber. Furthermore, properties of the material, necessary for such calculations (porosity and diffusion length), are often not known. In many cases, the change from the free emanation rate to the bound steady-state emanation rate (after closing the chamber) is so fast that it cannot be detected by measurement. These problems raise questions about the validity of the closed chamber method for determination of the free emanation rate.

Usual practice in measurements involving the closed chamber method is to assume that the volume of chamber is large relative to the pore volume of the sample to assure that free emanation takes place in the chamber. Another approach is to assume that at least at the beginning (after closing the chamber), when the radon concentration in the chamber is still very low, radon emanation may be treated as free emanation.[3,4] On the basis of this assumption, measurement of the initial radon growth rate in the chamber allows calculation of the free emanation rate.

The first of the previous two assumptions may not always be fulfilled, and the second may not always be justified. As shown by Samuelsson and Pettersson, the transition from free to bound emanation may occur within an hour or so after closing the chamber.[2] This means that the initial growth rate is not caused by free emanation.

The emanation rate determined on the basis of closed chamber measurements is the emanation rate occurring under constant pressure. In reality, all materials are affected by atmospheric pressure variations, which cause variations in the emanation rate. For this reason, instantaneous emanation rates may be different from emanation rates determined in the laboratory. Over the long term, however, the effects of atmospheric pressure variations usually cancel each other, and the emanation rate determined under stable conditions may be assumed to be the average value.

Despite its shortcomings, the closed chamber method is still a valuable tool for measurement of emanation rate. However, in view of the problems described above, it is necessary to establish and employ certain practical criteria for application of the method.

The objective of this chapter is to compare the available mathematical approaches to calculation of the free emanation rate from closed chamber experimental data and, on the basis of the comparison, to suggest the optimal method of correlating experimental data with theoretical calculations. Special attention is given to the method that takes into account the effect of reduction in driving force for radon emanation from the sample. A final outcome of the chapter is to present practical criteria for measurement of the emanation rate.

DIFFUSION AND EMANATION THEORY

The typical case of a porous slab sealed hermetically in an emanation chamber is taken as the example. For mathematical simplicity, the one-dimensional case of radon diffusion and emanation is discussed. If the slab is, for example, of cubic shape, it is assumed that emanation takes place only through one pair of opposite walls of the cube, the other walls being covered with sealant to prevent emanation. A second possible approach is to consider the slab to be infinite in two of its dimensions so that emanation in only one dimension need be analyzed. Mathematically, these cases are identical. Restricting the model to a one-dimensional case does not decrease its usefulness. Real constructional elements (e.g., walls) have two dimensions that are much larger than the third, and much larger than the radon diffusion length in the material, so that effectively the radon flow from the slab to the outside air may be considered to be through a pair of opposite planes.

The one-dimensional differential equation governing the processes of radon production, diffusion, and decay in the slab is:[1]

$$\frac{\partial C(x,t)}{\partial t} = f - \lambda C(x,t) + D_{eff}\frac{\partial^2 C(x,t)}{\partial x^2} \quad (1)$$

where

$C(x,t)$ = concentration of radon in the pore space of the sample
f = production of radon in the pore space
λ = decay constant of radon
D_{eff} = effective diffusion coefficient of radon in the material

The radon emanation rate E can be expressed as:

$$E(t) = -\delta D_{eff}\frac{\partial C(x,t)}{\partial x}\bigg|_{x=d} \quad (2)$$

where

d = half-thickness of the sample (the cartesian coordinate system has its origin in the middle of the sample)

and

δ = porosity of the sample material.

As a result of radon emanation from the sample, the radon concentration growth in the chamber is described by the equation:

$$\frac{dN(t)}{dt} = -\lambda N(t) + \frac{q(t)}{V} \quad (3)$$

where

$q(t)$ = $E(t)\cdot S$ is the total emanation rate of a sample (S is the surface area of the sample)
V = the volume of the chamber accessible for emanation (this is the volume of the chamber minus the volume of the sample)
$N(t)$ = radon concentration in the chamber at time t after closing the chamber.

The emanation chamber is assumed to be hermetically sealed. In practice, it is always necessary to check whether the chamber is hermetically sealed. If it is not possible to prevent a small leakage and if the leakage rate is constant, the solution may be modified by replacing the value λ by the sum $\lambda + \lambda_l$ in Equation 3 and in the boundary conditions of Equation 1, where λ_l is the escape constant of radon from the chamber.[4] Pressure in the chamber should be kept constant throughout the measurement.

If the sample emanates for a period of time much greater than the radon half life, steady-state is achieved ($\partial C/\partial t = 0$) and the solution is:

$$C(x) = -\frac{f/\lambda}{\cosh\beta}\left\{\frac{1}{1+\tanh\beta/\alpha\beta}\right\}\cosh\frac{x}{L} + \frac{f}{\lambda} \tag{4}$$

where

L = diffusion length of radon in the material ($L = \sqrt{D_{eff}/\lambda}$)
β = d/L (d is the half-thickness of the sample)
α = $V/\delta V_s$ (V_s is volume of the sample)

Substituting Equation 4 into Equation 2 and solving leads to the steady-state emanation rate:

$$E(\infty) = \frac{\delta L f(\tanh\beta)}{1+\tanh\beta/\alpha\beta} \tag{5}$$

If $\alpha \to \infty$, which corresponds to the situation that the volume of the air surrounding the slab is infinite relative to the pore volume of the slab, Equation 5 takes a simpler form, and the free emanation rate is given by:

$$E = \delta f L \tanh\beta \tag{6}$$

For simplicity, the free emanation rate is denoted by E, as it is independent of time.

The emanation rate cannot be measured directly and is calculated from the growth of the radon concentration in the chamber. After closing the chamber, the emanation rate changes from the free emanation rate given by Equation 6 (providing that, before closing the chamber, steady-state free emanation conditions were present) to the bound emanation rate given by Equation 5, based on actual values of V and V_s. The values of E(t) — between E and E(∞) — can be found by substituting the full time-dependent solution of Equation 1 into Equation 2, and solving. The time-dependent solution of Equation 1 is complicated and has been given by Krisiuk et al. and Samuelsson and Pettersson.[1,2] Equation 3 should be solved with the explicit form of E(t). The final form of N(t) should allow calculation of one of the following parameters: L, δ, η (emanation coefficient), if all required parameters are known. Alternatively, the complete set of parameters could be fitted into the experimental data. In practice, however, neither procedure is usually possible, because all required parameters are usually not known. The complicated form of the expression N(t) also makes fitting very difficult.

The dependences of E(t) and N(t) were presented graphically by Samuelsson and Pettersson, Samuelsson, and Samuelsson and Erlandsson for various values of L, δ, and η.[2,5,6]

However, from a practical point of view, knowledge of the change in the emanation rate with time is not necessary, because the change in the emanation rate characterizes the entire system, not just the sample. The sample is characterized by its free emanation rate.

Difficulties in correlating experimental data of radon concentration growth in the chamber with the free emanation rate are usually overcome by two approaches. The first approach is to assume that the value of α is large enough that the emanation in the chamber may be approximated with negligible error by the free emanation rate. In this case, solution of Equation 3 takes the simple form:

$$N(t) = \frac{q}{\lambda V}\left(1 - \exp(-\lambda t)\right) \tag{7}$$

From Equation 7, on the basis of experimental data, it is easy to find the constant value of q, the total free emanation rate. However, in practice, the value of α is not always large enough to allow free emanation to be approximated in the chamber. A conflicting requirement is that for material of a relatively low emanation rate, the volume into which emanation takes place should be small in order to achieve good counting statistics.

The second approach is to assume that, at least in the beginning, soon after closing the chamber, when the radon concentration in the chamber is still low, the emanation may be treated as free emanation. In this case, from the slope of the initial growth rate of the radon concentration in the chamber, the free emanation may be calculated from

$$r = \frac{N(t)}{t} = \frac{q}{V} \tag{8}$$

This approach, however, introduces several controversies. As shown by Samuelsson and Pettersson, the very period after closing the chamber is characterized by the fastest change in the emanation rate.[2] For example, for $V = V_s$ and for L = 100 cm, the emanation rate reaches a bound steady-state value within 4 h after closing the chamber. The ratio of bound to free emanation rates for these conditions is 0.65. It is obvious that, in this case, the slope of the initial growth rate would not provide the value of the free emanation rate. Sometimes, however, estimation of the free emanation rate from the initial growth rate may be possible. This may be true when, because of the geometry of the measurement, emanation is at all times close to free emanation and, for small values of t ($t < 24$ h), the exponent in Equation 7 may be approximated by the two first terms of the series:

$$\exp(-\lambda t) = 1 + \frac{-\lambda t}{1!} + \frac{(-\lambda t)^2}{2!} + \frac{(-\lambda t)^3}{3!} + \ldots \approx 1 - \lambda t, \tag{9}$$

Then Equation 7 takes the form:

$$N(t) = \frac{q}{\lambda V}(1 - 1 + \lambda t) = \frac{qt}{V} \tag{10}$$

where q is the total free emanation rate.

In applying this method to determine the free emanation rate, important assumptions are that the emanation rate has a constant value during the course of the measurement, and that the exponential growth of radon activity in the chamber is approximately linear. Assuming a constant value of the emanation rate at a time when the emanation rate changes the most rapidly is synonymous with assuming that free emanation takes place in the chamber all of the time.

In summary of the previous discussion, the two commonly used methods for determining the free emanation rate may be applied only when free emanation takes place in the chamber all of the time. The emanation rate may be determined (1) from Equation 7 (for example, by fitting experimental data to the exponential curve given by this equation), or (2) from Equation 10. The advantage of method (1) is that the time of the measurement is not restricted to the period after closing the chamber, which is important in the case of samples of low emanation rate. For such samples, the radon concentration in the chamber within the first few hours after closing the chamber may not be high enough to allow reasonable counting statistics. Use of method (2) allows the time of measurement to be shortened significantly. Method (2) can be used for samples having a relatively high emanation rate.

Another approach to the problem of free emanation rate measurements was presented by Wojcik and Morawska.[7] The model developed does not focus on the processes taking part inside the sample, but rather on their influence on growth of the radon concentration in the chamber. The principal assumption of this model is that if the volume of the chamber is not considerably greater than the volume of the sample, the radon present in the chamber reduces the driving force for emanation from the sample. The smaller the volume of the emanation chamber relative to the sample, the greater is the effect of the driving force reduction. The radon concentration vs. time relationship in the chamber is given by the following equation in which the effect of the reduction in driving force is taken into account:

$$\frac{dN(t)}{dt} = -\lambda N(t) + \frac{q}{V}\left(1 - \frac{N(t)}{N_p}\right) \tag{11}$$

In this equation, N_p is the maximum radon concentration in the pore space of the sample material, whose value depends on the radium concentration in the material, the porosity of the material, and the radon emanation coefficient. The maximum radon concentration in the pore space of the sample depends only on the condition of the sample at the time of the measurement (the moisture content, which affects the emanation coefficient) and not on the way in which the measurement is performed. The maximum concentration is reached when the difference between the volume of the chamber and the volume of the sample is much smaller than the volume of pore space within the sample. The reduced driving force effect described here may be compared to the term back diffusion discussed with different meanings, by Jonassen and Samuelsson and Pettersson.[4,6]

The second term on the right-hand side of Equation 11 may be compared to the second term on the right-hand side of Equation 3 to give:

$$E(t) = E\left(1 - \frac{N(t)}{N_p}\right) \tag{12}$$

which may be considered to be an approximation of the time-dependent emanation rate.

The solution to Equation 11 is:

$$N(t) = \frac{q}{V\left(\lambda + q/VN_p\right)}\left\{1 - \exp\left[-t\left(\lambda + \frac{q}{VN_p}\right)\right]\right\} \tag{13}$$

Thus it is possible to determine the total free emanation rate, q, of a sample, and the maximum radon concentration in the pore space of the sample, N_p, by making a time series of measurements of radon concentration growth in the chamber and applying Equation 13. The advantages of this method are that it is not necessary to assume that emanation in the chamber is free (and it does not have to be free) and that knowledge of the internal parameters of the sample (δ and L) is not necessary in order to determine the free emanation rate from the experimental data.

The validity of the results obtained in this way may be verified via comparison with theoretical calculations performed on the basis of Equation 5. As a basis for these comparisons, an experimental set of data may be used.[8] A sample of lightweight concrete of volume 3.375×10^{-3} m^3 was enclosed hermetically in a chamber of volume 3.95×10^{-2} m^3. The porosity of the material was 0.0733. The values obtained for the total free emanation rate and the maximum concentration of radon in the pore space of the material by fitting the time series of radon concentration in the chamber to Equation 13 are $q = 4.11 \times 10^{-5}$ Bq s^{-1} (19.6 atom s^{-1}), and $N_p = 69.7 \times 10^3$ Bq m^{-3}. The half-thickness of the sample was much smaller than the diffusion length, so that the ratio tanh β/β may be assumed to be unity.

According to Equation 12, the ratio of the bound steady-state emanation rate to the free emanation rate may be expressed as:

$$\frac{E(\infty)}{E} = 1 - \frac{N(\infty)}{N_p} \tag{14}$$

Table 1. Ratio of the Bound Steady-State to the Free Emanation Rate Calculated from Experimental Data and Theoretical Model as a Function of the Volume of the Chamber

$V[M^3]$	$\frac{V}{V_s}$	$\alpha = \frac{V^a}{\delta V_s}$	$\frac{E(\infty)}{E} = 1 - \frac{N(\infty)}{N_p}$	$\frac{E(\infty)}{E} = \frac{1}{1 + 1/\alpha}$
0.0361	10.67	145.9175	0.9928	0.9932
0.0200	5.93	80.8407	0.9861	0.9878
0.0100	2.96	41.4204	0.9725	0.9758
0.0050	1.48	20.2102	0.9465	0.9528
0.0010	0.30	4.0420	0.7795	0.8017

[a] In this case, $dV_s = 0.2474 \times 10^{-3} m^3$

The values of this ratio were calculated for different values of the volume of the chamber for the values q and N_p given above ($q(\infty)/q = E(\infty)/E$). The results are shown in the fourth column of Table 1. The same ratio, calculated from the theoretical Equations 5 and 6, is given by:

$$\frac{E(\infty)}{E} = \frac{1}{1 + \tanh \beta / \beta \alpha} \tag{15}$$

Values of this ratio were calculated for different values of a, assuming that tanh b/b is unity. The results are shown in the fifth column of Table 1.

From Table 1, it can be seen that values of the ratio of the bound steady-state emanation rate to the free emanation rate calculated from the experimental data and from the theoretical model are in excellent agreement. Small differences may be due to errors in porosity determination or to the fact that the ratio tanh b/b might not be exactly equal to unity.

As a summary of this discussion, it may be concluded that the free emanation rate can be determined by the closed chamber method using Equation 13 independent of the geometry of the measurement, that is, independent of whether free or bound emanation takes place in the chamber. This method is the simplest for determining the free emanation rate from experimental data without using complicated theoretical expressions. The result is the same as would be obtained by exact theory.

The values of time-dependent emanation rate, calculated accordingly to Equation 13 for small values of t, differ slightly from those calculated from Equation 2, but approach the same final value.[5] The change from free to bound emanation is faster according to the theoretical model described by Equation 2. However, the theoretical calculation is based on simplifying assumptions, for example, the assumption of instantaneous air mixing in the outer volume, which would cause the emanation process to be more rapid. As mentioned previously, experimental emanation rate vs. time data are not available, because the emanation rate is not measured directly.

EXPERIMENTAL DETERMINATION OF RADON EMANATION PROPERTIES

Experimental methods for applying the closed chamber method to determine radon emanation properties will now be considered. The radon emanation properties are the total emanation rate of a sample, the specific mass emanation rate, the specific surface emanation rate, and the emanation coefficient.

Total Free Emanation Rate of a Sample

In planning to measure the total free emanation rate of a sample, it is necessary to take into account the properties of the sample as well as the following requirements for the experiments:

- A rough estimate of the order of magnitude of the emanation rate, on the basis of the radium content of the sample. This estimate allows selection of an appropriate sample mass for reasonable counting statistics (for example, several grams for uranium ore, but several kilograms for bricks).
- Estimation of the radon diffusion length and the porosity of the material. Usually, the porosity of materials such as sand, silt, or clay is between 0.4 and 0.6, and the diffusion length in the range 100 to 200 cm for dry materials, and of the order of 1 to 3 cm for materials saturated with water. Both the porosity and the diffusion lengths of brick and concrete are smaller: the porosity of concrete is around 0.1 (bricks may be higher) and the dry diffusion length is usually not greater than 50 cm.[9] Knowledge of porosity and diffusion length is helpful in finding the optimal relationship between the dimensions of the sample and the volume of the emanation chamber.
- The duration of the experiment. The duration of the experiment may be decreased significantly by determining the total free emanation rate of a sample from the initial slope of the radon concentration vs. time curve; however, in this case, it is necessary to arrange conditions such that radon emanation may be considered "free" throughout the measurements. Results for the ratio $E(\infty)/E$ (determined from Equation 15) are shown in Table 2a—d for different values of diffusion length and porosity as a function of the volume of the material V_s and the volume of the emanation chamber, V. Tables 2a and 2b give data for material similar to typical dry and wet soil, respectively (porosity and diffusion length large, and porosity high and diffusion length small, respectively). Results for material similar to typical dry and wet concrete are given in Tables 2c and 2d.

  From Table 2, it is clear that for materials similar to dry soil, it is difficult to create a condition of free emanation in the chamber ($E(\infty)/E \approx 1.0$), except for very small samples in a very large chamber. The situation changes slightly for wet soils, but still the volume of the chamber must be 10 to 20 times larger than the volume of the sample in order to ensure free emanation for thick samples.

Table 2. The Ratio E (∞)/E Calculated for Different Values of Diffusion Length and Porosity as a Function of the Volume of the Material and the Volume of the Chamber

Table 2a

Diff. Length = 100.0 cm
Porosity = 0.500

	H[cm][a]						
d[cm]	1	5	10	20	50	80	100
1	0.7	0.9	1.0	1.0	1.0	1.0	1.0
5	0.3	0.7	0.8	0.9	1.0	1.0	1.0
10	0.2	0.5	0.7	0.8	0.9	0.9	1.0
20	0.1	0.3	0.5	0.7	0.8	0.9	0.9
50	0.9	0.2	0.3	0.5	0.7	0.8	0.8
80	0.0	0.1	0.2	0.4	0.6	0.7	0.8
100	0.0	0.1	0.2	0.3	0.6	0.7	0.7

Table 2b

Diff. Length = 2.0 cm
Porosity = 0.500

	H[cm][a]						
d[cm]	1	5	10	20	50	80	100
1	0.7	0.9	1.0	1.0	1.0	1.0	1.0
5	0.5	0.8	0.9	1.0	1.0	1.0	1.0
10	0.5	0.8	0.9	1.0	1.0	1.0	1.0
20	0.5	0.8	0.9	1.0	1.0	1.0	1.0
50	0.5	0.8	0.9	1.0	1.0	1.0	1.0
80	0.5	0.8	0.9	1.0	1.0	1.0	1.0
100	0.5	0.8	0.9	1.0	1.0	1.0	1.0

Table 2c

Diff. Length = 20.0 cm
Porosity = 0.050

	H[cm][a]						
d[cm]	1	5	10	20	50	80	100
1	1.0	1.0	1.0	1.0	1.0	1.0	1.0
5	0.8	1.0	1.0	1.0	1.0	1.0	1.0
10	0.7	0.9	1.0	1.0	1.0	1.0	1.0
20	0.6	0.9	0.9	1.0	1.0	1.0	1.0
50	0.5	0.8	0.9	1.0	1.0	1.0	1.0
80	0.5	0.8	0.9	1.0	1.0	1.0	1.0
100	0.5	0.8	0.9	1.0	1.0	1.0	1.0

Table 2. The Ratio E (∞)/E Calculated for Different Values of Diffusion Length and Porosity as a Function of the Volume of the Material and the Volume of the Chamber (continued)

Table 2d

Diff. Length = 2.0 cm
Porosity = 0.050

	H[cm][a]						
d[cm]	**1**	**5**	**10**	**20**	**50**	**80**	**100**
1	1.0	1.0	1.0	1.0	1.0	1.0	1.0
5	0.9	1.0	1.0	1.0	1.0	1.0	1.0
10	0.9	1.0	1.0	1.0	1.0	1.0	1.0
20	0.9	1.0	1.0	1.0	1.0	1.0	1.0
50	0.9	1.0	1.0	1.0	1.0	1.0	1.0
80	0.9	1.0	1.0	1.0	1.0	1.0	1.0
100	0.9	1.0	1.0	1.0	1.0	1.0	1.0

[a] For simplicity, the volume of the chamber is expressed as the height of the air column above the sample, H, and the volume of the sample is expressed as its half-thickness, d.

On the other hand, for materials of properties similar to typical wet concrete, the volume of the chamber need be only slightly higher than the volume of the sample in order to create the conditions for free emanation.

If the conditions for free emanation in the chamber are fulfilled, the total free emanation rate may be calculated from Equations 7 or 10. If these conditions are not fulfilled, or if there are doubts as to whether the emanation is free, Equation 13 should be applied.

Specific Mass Emanation Rate

The specific mass emanation rate [$Bq\ kg^{-1}s^{-1}$] may be calculated directly from the total free emanation rate when the sample is so thin relative to the diffusion length (small value of β) that the ratio $\tanh \beta/\beta$ may be assumed to be unity. Under these conditions, essentially all radon that enters the pore space of the sample will be able to emanate from the sample to the chamber. However, escape of all radon entering the pore space may not occur for all cases of free emanation in the chamber even when E(∞)/E is close to unity, because it may be that when $\tanh \beta/\beta < 1.0$, $1/\alpha \ll 1.0$ (see Equation 15).

The value of $\tanh \beta/\beta$ may be calculated on the basis of known values of the diffusion length. It is easy to check experimentally whether the assumption that $\tanh \beta/\beta \ll 1.0$ is justified. In order to do this, a second sample of smaller dimensions should be measured. If the determined values of the mass emanation rate are equal for the original sample and for the smaller sample, the implication is that for both cases $\tanh \beta/\beta$ equals unity.

Specific Surface Emanation Rate

The specific surface emanation rate [Bq $m^{-2}s^{-1}$] is usually defined as the surface emanation rate of thick samples when further increase in the thickness does not alter the radon flux through the sample surface. In this case, the closed-chamber method can be applied by sealing the chamber to the sample surface, instead of enclosing the sample in the chamber. The surface emanation rate may be calculated from Equation 5 or, for free emanation conditions, from Equation 6. With the increasing β (increasing thickness of the sample) tanh $\beta \rightarrow 1.0$, and the surface emanation rate tends to a constant value given by $E = \delta fL$.

For example, if the half-thickness of the sample is twice the diffusion length ($d = 2L$, as recommended by Jonassen 1983), the value of tanh β is 0.964. In order to satisfy the condition tanh $\beta \approx 1.0$, the thickness of a concrete sample should be of the order to 1 to 2 m (if the diffusion length is in the range 0.25 to 0.5 m), and of a soil sample (which has a higher diffusion length), of the order of several meters. Such large samples usually cannot be enclosed in an emanation chamber unless the dimensions of the chamber are very large. Instead, for practical reasons, it is better to seal a smaller chamber to the surface of the material.

Emanation Coefficient

The emanation coefficient may be calculated from the value of the total free emanation rate (independent of whether determined from the conditions of free or bound emanation in the chamber), but only when tanh $\beta/\beta = 1.0$. The emanation coefficient is given by the equation:

$$\eta = \frac{q}{C_{Ra} m} \tag{16}$$

where C_{Ra} is the radium concentration of the sample (units of Bq kg^{-1}) and m is the mass of the sample.

Diffusion Length and Porosity

The closed-chamber method may allow determination of the porosity and diffusion length of the material, in addition to determination of the radon emanation properties. If emanation rate measurements are performed for two cases for which the values of the ratio $E(\infty)/E$ are different (< 1.0) (calculated from Equation 14), it is possible to calculate the values of L and δ from Equation 15. For example, for $d = 5$ cm and $H = 1$ cm (see Table 2a), $E(\infty)/E = 0.3$, and for $d = 5$ cm and $H = 10$ cm, the value of this ratio is 0.8. From Equation 15, values of α and β may be determined, and from these, L and δ.

There are, however, restrictions on application of the closed-chamber method for determining diffusion length and porosity because, for some materials, it is difficult to create conditions favoring distinctly different values of the ratio $E(\infty)/E$ (see, for example, Table 2d).

CONCLUSIONS

This chapter provides a detailed discussion of application of the closed-chamber method to determination of radon emanation properties. Despite the fact that the closed-chamber method in different forms has been used for decades, and the fact that theoretical models of the processes taking part in the sample and in the chamber have been developed, application of the method is still controversial, because it is very difficult to correlate experimental data with the complicated theoretical expressions in order to calculate the free emanation rate of the sample. Furthermore, such calculations are impossible unless the sample diffusion length and the sample porosity are known.

On the basis of the analysis provided here, the following conclusions may be drawn:

- The two common methods of determining the free emanation rate via the closed-chamber method, namely, (i) fitting experimental data to the exponential form of the radon concentration growth curve in the chamber, and (ii) determining the initial slope of the concentration growth curve, may be used only if the emanation in the chamber, because of the geometry of the measurement, can be treated as free emanation.
- If the requirements for free emanation cannot be fulfilled, or if it is difficult to judge whether emanation in the chamber is free or bound, it is suggested that for free emanation rate calculations, use be made of the expression taking into account the effect of the reduction in driving force caused by the radon present in the chamber. The form of this expression is simple (equation 13) and does not require knowledge of internal properties of the sample such as porosity or diffusion length. It was demonstrated here that values of the free emanation rate determined in this way are in excellent agreement with values calculated from the exact theoretical model.

The last section of this chapter identifies practical criteria for measuring emanation properties. The results presented for the ratio $E(\infty)/E$, for various values of diffusion length and porosity, as a function of pore volume of the sample and volume of the emanation chamber, are intended to be of practical use in planning emanation rate measurements, especially if there is a requirement for creating conditions of free emanation in the chamber.

Certain other parameters can be also calculated from the free emanation rate, namely, the specific mass emanation rate, the surface emanation rate, the emanation coefficient, the diffusion length, and the porosity. Conditions are discussed under which these parameters can be determined.

ACKNOWLEDGMENTS

This work was supported by a grant from the Natural Sciences and Engineering Research Council of Canada.

REFERENCES

1. Krisiuk, E.M., Tarasov, S.J., Shamov, V.P., Shalok, N.J., Lisachenko, E.P., and Gomelsky, L.G., "A study of radioactivity of building materials," (Leningrad: Ministry of Public Health of RSFSR, Research Institute for Radiation Hygiene, 1974).
2. Samuelsson, Ch. and Pettersson, H., "Exhalation of Radon-222 from Porous Materials," *Radiat. Prot. Dosimetry,* 7(1-4):95–100 (1983).
3. Jonassen, N. and McLaughlin, J.P., "Exhalation of Radon-222 from Building Materials and Walls," in Natural Radiation Environment III, U.S. Dept. of Energy CONF-780422, Gesell, T.F. and Lowder, W.M., (Eds.), 1211–1224, (1980).
4. Jonassen, N., "The Determination of Radon Exhalation Rates," *Health Phys.,* 52(2):369–376 (1983).
5. Samuelsson, Ch., "A Critical Assessment of Radon-222 Exhalation Measurements Using the Closed-Can Method," in *Radon and Its Decay Products: Occurrence, Properties and Health Effects,* Hopke, P.K., (Ed.), (Washington, DC: American Chemical Society, 1987).
6. Samuelsson, Ch. and Erlandsson, K., "The Implications of the Time-Dependent Diffusion Theory on Radon-222 Exhalation Measurement," (Sydney, Australia: 7th Annual Congress of the International Radiation Protection Association (IRPA-7), April 10–17, 1988).
7. Wojcik, M. and Morawska, L., "Radon Concentration and Exhalation Measurements with a Semiconductor Detector and an Electrostatic Precipitator Working in a Closed Circulation System," *Nucl. Instrum. Methods,* 212:393-402 (1983).
8. Morawska, L., "Two Ways of Determining the ^{222}Rn Emanation Coefficient," *Health Phys.,* 57(3):481–483 (1989).
9. Nazaroff, W.W. and Nero, A.V., Jr., *Radon and Its Decay Products in Indoor Air,* (New York: John Wiley & Sons, 1988).

CHAPTER 16

Determination of the Combined Collection-Counting Efficiency of Radon Progeny on Wire Screens

C.R. Phillips, L. Morawska, A. Khan, and I.-C. Hsieh

ABSTRACT

Practical application of wire screens to the determination of the unattached fraction of radon progeny or the size distribution of radon progeny requires knowledge of the combined collection-counting efficiency of the wire screen. The term unattached fraction is used here to describe those radon progeny that are not attached to atmospheric aerosols.

Although there have been several theoretical and experimental attempts to determine the collection efficiency and the combined collection-counting efficiency of wire screens, definitive data are still lacking. In this work, collection and counting data computed on the basis of new theoretical models are evaluated, together with experimental results obtained using two different methods. The theoretical calculations and the experimental measurements were made for seven different wire screen sizes (35, 40, 50, 60, 80, and 120 mesh Tyler screen) and three different face velocities (7.70, 10.41, and 13.18 cm s^{-1}). Recommended values of the combined collection-counting efficiencies are given for these screens.

INTRODUCTION

The wire screen method for measuring unattached radon or thoron progeny consists of drawing an air sample through a wire screen filter assembly, with the filter behind the screen. The unattached radon/thoron progeny atoms deposit on the wire screen because of their high diffusion coefficients. The wire screen is counted in an alpha counter and the unattached activity determined from the counts. Although this method appears to be simple and practical, it suffers from a lack of precision because neither the collection efficiency nor the counting efficiency of the wire screen is known precisely. The collection efficiency has been determined theoretically by Thomas and Hinchliffe based on a simplified model of the wire screen as an assembly of short cylindrical tubes, taking the thickness of the wire screen as the length of each cylinder.[1] Another simplified model was proposed by Cheng and Yeh in which the wire screen was modeled as a fibrous filter.[2] George determined experimentally the collection efficiency of a 60 mesh per inch Tyler wire screen.[3] He found "zero" efficiency for collection of activity attached to aerosols. At 11.5 cm s^{-1} face velocity, George found that the experimentally determined collection efficiency (0.77) agreed well with the Thomas and Hinchliffe equation (0.76). George presented no results for other mesh sizes or for an arbitrary flowrate.

The counting efficiency of a wire screen has been determined experimentally in other studies. Estimates of the geometric counting efficiency range from 0.70 to 1.00.[3-6]

DETERMINATION OF THE COMBINED COLLECTION-COUNTING EFFICIENCY OF RADON PROGENY ON A WIRE SCREEN

In the present work, the collection efficiency and counting efficiency were first determined theoretically for a 60-mesh Tyler wire screen. (Previously, only the experimental geometric efficiency has been determined.) The collection efficiency and the combined collection-counting efficiency were then determined experimentally. Measurement conditions were the same as used for the theoretical calculation. The wire screen was held in a standard 25 mm open-face filter holder. In the screen, the wire diameter was 0.178 mm and the size of each opening 0.25 mm. For the purpose of calculation, it was assumed that the wires butt and do not overlap. The effective diameter of the wire screen when mounted in a filter holder (with a 1-mm diameter O-ring) was 23 mm. Wire screen counting was carried out in a standard Trimet alpha counter (counting aperture, 50 mm diameter).

For the measurements, it was assumed that the unattached fraction consists only of free molecular ^{218}Po, not including ultrafine particles or clusters. In order to prevent cluster formation, the radon-carrying air was filtered (HEPA filter, followed by glass wool column and silica gel column) before entering the sampling container. The residence time of the air in the jar was also kept as short as possible. If the unattached fraction consists of clusters of diameter larger than molecular ^{218}Po (and therefore having a lower diffusion coefficient), the calculated collection efficiency will be overestimated. This problem was recently discussed by Ramamurthi and Hopke.[7] The aim of the present experimental work was to test the theoretical model under conditions of a molecular, unattached fraction in the sampling device. The model can be applied to a cluster spectrum (0.5 to 3 nm) by taking into account the actual properties of the spectrum.

Theoretical Determination of the Collection and Counting Efficiency of a Wire Screen

Procedure for Determination of Counting Efficiency

For the purpose of calculation, the wire screen was considered to be a stack of horizontal planes (thickness of each plane = 0.001 mm). The opening size varies for each plane, being determined by the separation between the wire surfaces at each particular point along the axis of air flow. The total deposition on the wire screen is then given by the combined deposition on the sides of all openings for all 178 planes.

The following conditions were applied for the calculations:

1. Activity is deposited uniformly around each wire.
2. Because of symmetry, the view factor of a quarter of the screen is used to determine the view factor for the entire screen.
3. The distance between the detector and the first plane of the wire screen is 1 mm.
4. Neighboring parallel wires only are tested to check whether they block passage of the alpha particles to the detector.
5. The center of the cartesian coordinate system is chosen to be at the center of the intersection of the central wires in the x and y directions.

The calculation procedure is based upon taking one horizontal plane of 0.001 mm thickness at a time and choosing points along the x-axis at a specified step size (0.004 mm) until the edge of the wire screen is reached. For determining the x- and y-coordinates at every point, ΔH shown in Figure 1 is calculated in order to take the curvature of the wire into account:

$$r = L + \Delta H \tag{1}$$

$$\Delta H = r - L = r - \sqrt{\left[r^2 - \left(r - (D - 1)\right)^2\right]} \tag{2}$$

where

r = the wire radius
D = the vertical distance between the detector plane and the plane under consideration
L = the horizontal distance between the point under consideration and the z-axis.

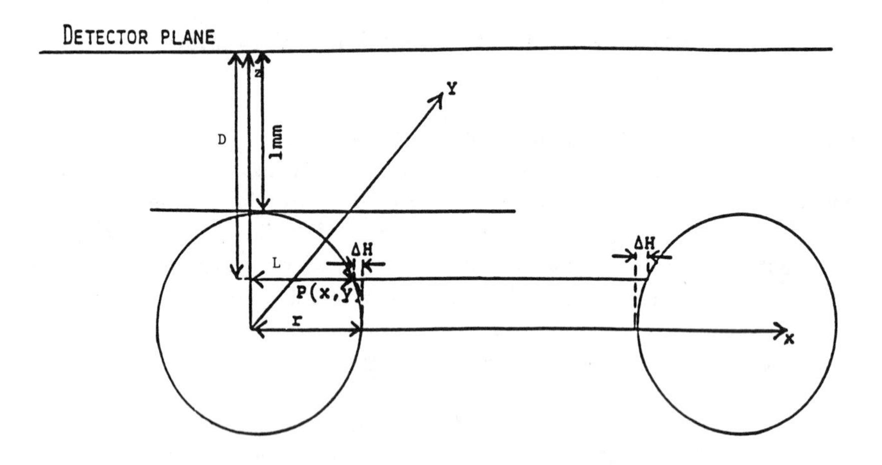

Figure 1. Choosing the x- and y-coordinates using ΔH.

At every plane, the starting point of the calculation at the center of the wire screen is then given by the following x- and y-coordinates:

$$x = r - \Delta H \tag{3}$$

$$y = r - \Delta H \tag{4}$$

The number of iterations required depends upon the width of each unit cell, which is given by:

$$\text{width} = 2\Delta H + 0.678$$

For a step size of 0.004 mm, the number of iterations is given by (width)/(0.004).

At each point, the computer program uses a random number generator to determine the direction cosine of a disintegration emitted in a random direction. The random direction is, however, limited to a cone with solid angle $4\pi\eta$ that encompasses the detector. η is the angle subtended by the detector at the wire screen plane under consideration.

Interference from a neighboring wire is determined in the following manner. For a point located in the upper half of the wire, Figure 2a shows how the critical angle, θ_{cri}, is calculated. θ_{ran} indicates the random direction of emission of an alpha disintegration. In order that the emitted alpha misses the neighboring wire, θ_{ran} must be less than or equal to θ_{cri}, which is the angle formed by the tangent to the neighboring wire at the point under consideration and the z-axis. From Figure 2a it can be seen that:

$$\begin{aligned}\theta_{cri} &= \pi/2 - (\theta_1 - \theta_2) \\ &= \pi/2 - \left(\sin^{-1}(r/AX)\right) - \sin^{-1}\left((r - (D-1)/AX)\right)\end{aligned} \tag{5}$$

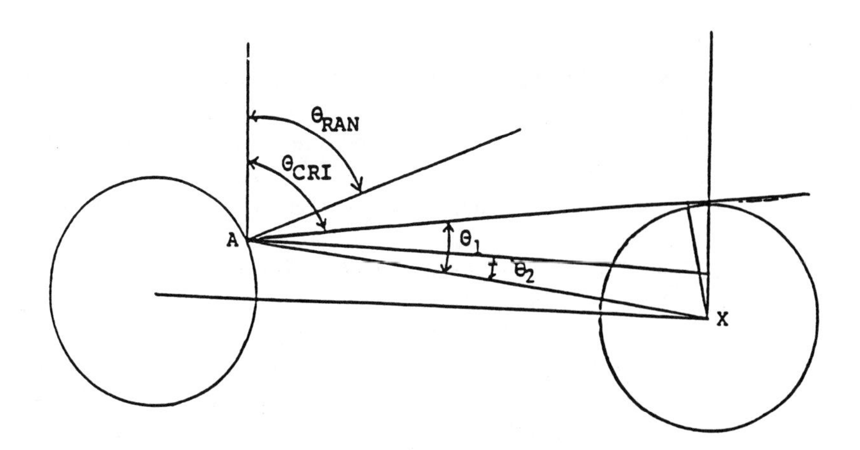

Figure 2a. Critical angle calculation for a point in the upper half of the wire.

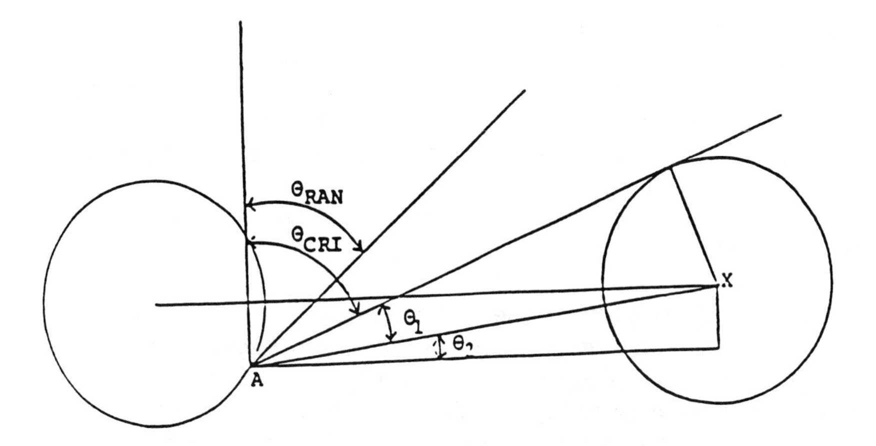

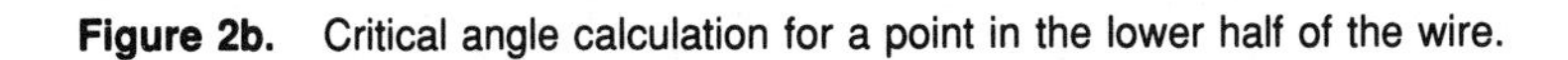

Figure 2b. Critical angle calculation for a point in the lower half of the wire.

Similarly, for a point located in the lower half of the wire, Figure 2b shows that:

$$\begin{aligned}\theta_{cri} &= \pi/2 - (\theta_1 + \theta_2) \\ &= \pi/2 - \left(\sin^{-1}(r/AX) + \sin^{-1}\left(((D-1)-r)\right)/AX\right)\end{aligned} \tag{6}$$

A check is made to determine whether the point at which the emitted alpha is incident on the detector plane lies within or outside the detector.

If it is found that the emitted alpha does not hit the neighboring wire and, thus, reaches the detector, the event is recorded as a hit. In the next iteration, a new point is chosen along the x-axis at a distance of 0.004 mm from the previous point, and the entire procedure, described earlier, of determining whether an alpha disintegration emitted from this point will hit the detector or miss the detector, is repeated. At the intersection of a wire, the step size is increased in order to skip the intersecting wire. When the end of the wire screen is reached, the next point is chosen along the next wire at the center of the screen and the y-coordinate is increased by one width of the unit cell (0.428 mm). The procedure is repeated until the edge of the wire screen is reached in the y-direction. The program then moves on to the next plane.

After calculations have been completed for all the 89 planes for each half of the wire screen, the geometric counting efficiency (view factor) of the wire screen is calculated as $\eta \times$ (total number of hits/total number of events).

The view factors for the upper and lower halves of the screen are calculated separately because there are minor differences in the equations for calculating θ_{cri} for points in the upper half and the lower half of the screen. An average for the view factors for the two halves is then taken as the final view factor for the wire screen.

Results for Counting Efficiency

Using 178 planes, each 0.001 mm thick, and x-axis steps of 0.004 mm, the view factor for the upper half of the screen was calculated as 0.467. Similarly, the view factor for the lower half was calculated as 0.344. The average view factor for the screen is therefore (0.467 + 0.344)/2 = 0.405. This view factor is relative to a view factor of 0.5 for a filter. If the geometric counting efficiency of a filter is represented as 1.0, the wire screen geometric efficiency, according to the present work, becomes 0.81.

Theoretical Calculation of Collection Efficiency

The simplest wire screen model was used for preliminary calculations of collection efficiency. The wire screen opening was assumed to be a square opening with the inner surface stretched flat (Figure 3). Based on a 60 mesh screen, the width of an opening was taken to be 0.0340 cm (measured between two points each midway between the wire center and its edge (Figure 3)) and the length of the opening to be 0.0280 cm (half the circumference of the wire).

In order to model the diffusional deposition of unattached radon-thoron progeny on the wire screen at a pump flowrate of 2 L min^{-1} (face velocity = 11.5 cm s^{-1}), the steady-state diffusion equation in rectangular coordinates for laminar flow in three dimensions was used as the starting point:

$$V_x \frac{\partial C}{\partial X} + V_y \frac{\partial C}{\partial Y} + V_z \frac{\partial C}{\partial Z} = D\left(\frac{\partial^2 C}{\partial X^2} + \frac{\partial^2 C}{\partial Y^2} + \frac{\partial^2 C}{\partial Z^2}\right) \quad (7)$$

where V_x, V_y, and V_z are the velocity components in the X, Y, and Z directions, respectively, C is the concentration of unattached radon/thoron progeny, and D is the diffusivity of unattached radon/thoron progeny.

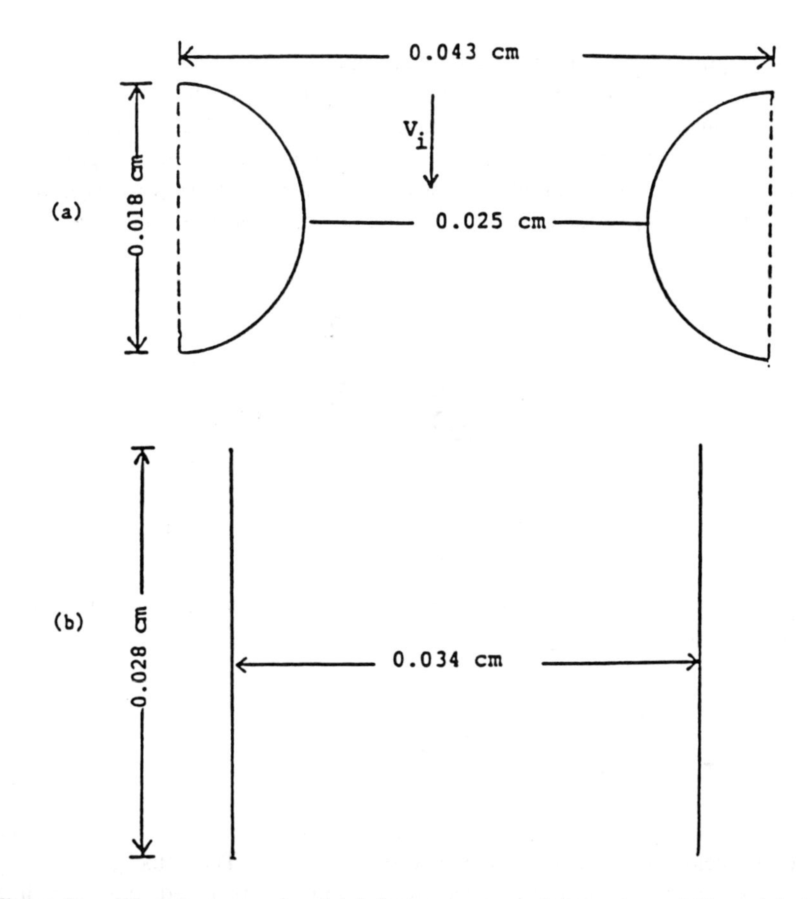

Figure 3. Wire screen opening. (a) Actual geometry and dimensions of 60 mesh/inch Tyler screen. (b) Simplified geometry.

This differential equation defines an initial-value problem with air flow along the z-axis. With initial and boundary conditions, the equation describes how C(x,y,z) propagates forward along the z-axis. As the initial condition, the concentration in the air before arriving at the front of the wire screen is assumed to be uniform. The boundary conditions, as Z increases, are Dirichlet conditions that specify the values of the boundary points. In the case of the wire screen, the concentrations on the wire surface are zero.

There are many finite-differencing schemes for diffusive initial-value problems. The main concern in choosing one is the computational stability of the algorithm. The Crank-Nicholson method is used in this calculation because it is of second-order accuracy and is usually stable for a large step forward.[8,9] Another method worth considering is the Alternating-Direction Implicit (ADI) method.[9,10] van der Vooren and Phillips developed a Modified Alternating-Direction Implicit (MADI) method for convective diffusion with a parabolic velocity profile in the more straight forward case of flow between parallel planes.[11] The present case of flow through a square opening requires a simplified approach.

If the superficial velocity is denoted by V_s and the interstitial velocity is denoted by V_i, then V_i is maximum at the equator of each wire, expressed as V_i(max) and V_i is minimum at the poles of each wire, expressed as V_i(min).

For the calculation, the interstitial velocity is used. For the geometry in Figure 3a,

$$V_i(\text{min}) = V_S = 11.5 \text{ cm s}^{-1} \tag{11}$$

$$V_i(\text{max}) = \frac{(11.5)(0.043)(0.043)}{(0.025)(0.025)} = 34.02 \text{ cm s}^{-1} \tag{12}$$

$$V_i(\text{avg}) = \frac{\left(V_i(\text{min}) + V_i(\text{max})\right)}{2} = 22.76 \text{ cm s}^{-1} \tag{13}$$

It is assumed that molecular diffusion in the direction of air flow is negligible compared to mass transport by convection, that is,

$$D\frac{\partial^2 C}{\partial Z^2} \to 0 \tag{14}$$

The total velocity of approach of air to the screen is based on the total screen area; this velocity, therefore, is the superficial velocity. The velocity within the screen openings (the interstitial velocity) is higher. As noted, the interstitial velocity is used. For a uniform velocity distribution within the screen opening, $V_x = V_y = 0$, and V_z = constant. The simplified convection diffusion equation is then

$$V_2 \frac{\partial C}{\partial Z} = D\left(\frac{\partial^2 C}{\partial X^2} + \frac{\partial^2 C}{\partial Y^2}\right) \tag{15}$$

This equation can be solved as a two-dimensional initial-value problem.

The Crank-Nicholson differencing scheme is applied in a two-dimensional form, with the following definitions of C:

$$C_{1_{(I,J)}} = \text{concentration at a grid point on a plane with coordinate Z} \tag{16}$$

$$C_{2_{(I,J)}} = \text{concentrationation at a grid po int on a plane with coordinate } Z + dZ \tag{17}$$

$$\frac{\partial C}{\partial Z} = \frac{C_{2_{(I,J)}} - C_{1_{(I,J)}}}{\Delta Z} \tag{18}$$

$$\frac{\partial^2 C}{\partial X^2} = \left(\frac{1}{2}\right)\left(\frac{C_{2_{(I+1,J)}} - 2C_{2_{(I,J)}} + C_{2_{(I-1,J)}}}{(\Delta X)^2} + \frac{C_{1_{(I-1,J)}} - 2_{1(I,J)} + C_{1_{(I-J)}}}{(\Delta X)^2}\right) \tag{19}$$

$$\frac{\partial^2 C}{\partial Y^2} = \left(\frac{1}{2}\right)\left(\frac{C_{2_{(I,J+1)}} - 2C_{2_{(I,J)}} + C_{2_{(I,J-1)}}}{(\Delta Y)^2} + \frac{C_{1_{(I,J=1)}} - 2C_{1_{(I,J)}} + C_{1_{(I,J-1)}}}{(\Delta Y)^2}\right) \tag{20}$$

On substituting these terms into the differential equation, the difference equation is obtained as

$$V_Z \frac{C_{2_{(I,J)}}}{\Delta Z} + D\frac{C_{2_{(I,J)}}}{\Delta X} + D\frac{C_{2_{(I,J)}}}{\Delta Y}$$

$$= V_Z \frac{C_{1_{(I,J)}}}{\Delta Z} + D\frac{C_{2_{(I+1,J)}} + C_{2_{(I-1,J)}} + C_{1_{(I+1,J)}} - 2C_{1_{(I,J)}} + C_{2_{(I-1,J)}}}{2(\Delta X)^2}$$

$$+D\frac{C_{2_{(I,J+1)}} + C_{2_{(I,J-1)}} + C_{1_{(I,J+1)}} - 2C_{1_{(I,J)}} + C_{1_{(I,J-1)}}}{2(\Delta Y)^2} \tag{21}$$

This equation was used to solve for $C_{2(I,J)}$ by Gauss-Seidel iteration. The following parameters were used in the initial computation:

- diffusion coefficient: as specified in data tables following
- initial particle concentration: 326 atoms L^{-1} (for 1 Working Level, ^{218}Po concentration at equilibrium is 977 atoms L^{-1}; at typical third of Working Level, concentration of unattached ^{218}Po is therefore 326 atoms L^{-1}
- step size in X, Y direction: 0.001 cm
- step size in Z direction: 0.0002 cm
- velocity in Z direction: as specified in data tables

The proportion of particles depositing on the inner side walls of the opening was calculated after each increment of Z coordinate by computing the average remaining particle concentration in the air.

Table 1. Deposition Efficiencies at a Fixed Interstitial Velocity (11.5 cm s^{-1}) and Varying Diffusivities for 60 Mesh Wire Screen Calculations with V_z = 11.5 cm s^{-1}

Diffusivity	Total Deposition (%)					
cm² s^{-1}	Step 1	Step 2	Step 3	Step 4	Step 5	Step 140
0.060	9.00	13.89	17.85	21.12	23.97	94.48
0.054	8.30	12.98	16.74	19.86	22.59	92.90
0.045	7.18	11.51	14.94	17.81	20.33	89.59
0.040	6.53	10.62	13.85	16.58	18.96	87.15
0.030	5.15	8.66	11.46	13.83	15.91	80.36
0.020	3.63	6.38	8.62	10.55	12.26	69.92
0.010	1.94	3.60	5.05	6.35	7.52	52.50
0.0000001	0.0	0.0	0.0	0.0	0.0	0.0
	Deposition in Each Step (%)					
0.060	9.00	4.89	3.96	3.27	2.85	
0.054	8.30	4.68	3.76	3.12	2.73	
0.045	7.18	4.33	3.43	2.87	2.52	
0.040	6.53	4.09	3.23	2.73	2.38	
0.030	5.15	3.51	2.80	2.37	2.08	
0.020	3.63	2.75	2.24	1.93	1.71	
0.010	1.94	1.66	1.45	1.30	1.17	
0.0000001	0.0	0.0	0.0	0.0	0.0	

Results for Deposition Efficiency Calculations

Calculated results for the 60 mesh wire screen for face velocities 7.70, 10.41, and 13.18 cm/s are 0.88, 0.81, and 0.75, respectively. Results of stepwise deposition for various diffusivities are shown in Tables 1 to 3.

Discussion

The geometric counting efficiency (view factor) of the 60 mesh wire screen was calculated as 0.81. The collection efficiency for the unattached radon/thoron progeny of diffusivity 0.054 cm²/s was found to be 0.80 for a face velocity of 11.5 cm/s. The overall efficiency of the 60 mesh wire screen for unattached radon/thoron progeny determination is, therefore, the product $0.80 \times 0.81 = 0.65$. It is important to keep the assumption of uniform deposition around the wires in perspective while interpreting the view factor result. The deposition, in fact, has a profile with gradually decreasing deposition in the direction of flow. This profile is supported by the experimental results of Holub and Knutson who found that the front-to-back ratio for α-counts for a 60 mesh screen was about 2.5 for a face velocity of 18 to 21 cm/s.[12] The deposition profile calculated as a part of the stepwise deposition calculation in the direction of air flow confirms this behavior. In Table 1, the stepwise deposition for a diffusivity of 0.06 cm²/s is 9% in the first step, 4.89% in the second step, 3.96% in the third step, and 3.27% in the fourth

Table 2. Deposition Efficiencies at a Fixed Diffusivity and Varying Interstitial Velocities for 60 Mesh Wire Screen Calculations with Diffusivity = 0.054 $cm^2\ s^{-1}$

Interstitial Velocity ($cm\ s^{-1}$)	Total Deposition (%)					
	Step 1	Step 2	Step 3	Step 4	Step 5	Step 140
34.02	3.35	5.93	8.06	9.90	11.54	67.61
23.0	4.71	8.02	10.66	12.91	14.90	77.76
22.76	4.75	8.08	10.74	13.00	15.00	78.02
18.04	5.68	9.43	12.39	14.91	17.11	83.31
11.5	8.30	12.98	16.74	19.86	22.59	92.90
	Deposition in Each Step (%)					
34.02	3.35	2.58	2.13	1.84	1.64	
23.0	4.71	3.31	2.64	2.25	1.99	
22.76	4.75	3.33	2.66	2.26	2.00	
18.4	5.68	3.75	2.96	2.52	2.20	
11.5	8.30	4.68	3.76	3.12	2.73	

Table 3. Deposition Efficiencies at a Fixed Interstitial Velocity (22.76 $cm\ s^{-1}$) and Varying Diffusivities for 60 Mesh Wire Screen Calculations with V_z = 22.76 $cm\ s^{-1}$ (Interstitial)

Diffusivity $cm^2\ s^{-1}$	Total Deposition (%)					
	Step 1	Step 2	Step 3	Step 4	Step 5	Step 140
0.060	5.20	8.73	11.54	13.92	16.02	80.64
0.054	4.75	8.08	10.74	13.00	15.00	78.02
0.045	4.07	7.05	9.46	11.52	13.35	73.25
0.040	3.67	6.43	8.69	10.63	12.35	70.16
0.030	2.84	5.11	7.01	8.67	10.16	62.64
0.020	1.96	3.63	5.09	6.40	7.58	52.76
0.010	1.02	1.96	2.82	3.63	4.38	37.99
0.0000001	0.0	0.0	0.0	0.0	0.0	0.0
	Deposition in Each Step (%)					
0.060	5.20	3.53	2.81	2.38	2.10	
0.054	4.75	3.33	2.66	2.26	2.00	
0.045	4.07	2.98	2.41	2.06	1.83	
0.040	3.67	2.76	2.26	1.94	1.72	
0.030	2.84	2.27	1.90	1.66	1.49	
0.020	1.96	1.67	1.46	1.31	1.18	

Table 4. **Comparison of Collection Efficiencies for 60 Mesh Wire Screen for Face Velocity V = 11.5 cm s^{-1}**

Diffusivity $cm^2\ s^{-1}$	Efficiency (%) by Thomas and Hinchliffe	Efficiency (%) by Cheng and Yeh	Efficiency (%) by Finite-Difference
0.060	0.754	0.813	0.806
0.054	0.723	0.790	0.780
0.045	0.668	0.749	0.733
0.040	0.633	0.722	0.702
0.030	0.551	0.652	0.626
0.020	0.451	0.553	0.528
0.010	0.329	0.398	0.380

step. The stepwise deposition pattern is similar for other values of diffusivity. A similar stepwise deposition pattern (Table 2) is also found for other interstitial velocities. The upper half of the screen, in practice, has a higher deposition density than the lower half. Since the lower half contributes less to the overall view factor, it can be concluded that the counting efficiency computed in the present work is a slight underestimation.

Interpretation of the collection efficiency results should be made carefully because of the simplified square cell model of the wire screen openings used for these calculations. The present results are compared with those of Thomas and Hinchliffe and Cheng and Yeh in Table 4.[1,2]

Ideally, the deposition profile for the actual wire screen geometry, consisting of wires running over and under, should be determined. The real wire screen geometry is so complex, however, that simplified models need to be constructed. The actual deposition profile should ideally be incorporated in the geometric counting efficiency calculations in order to obtain the true geometric efficiency of a wire screen for use in determining the unattached fraction of radon/thoron progeny.

Experimental Determination of Combined Collection-Counting Efficiency of a Wire Screen

Measurement Procedures

Several procedures may be used in the experimental determination of the collection and counting efficiency of a wire screen. The common basis for the procedures is simultaneous sampling through a reference filter and a wire screen with a back-up filter held behind the screen. Efficiency determination can be based on one of the following procedures:

1. Counting the reference filter and the back-up filter. This procedure yields information about the collection efficiency of the wire screen provided that

the unattached fraction of the radon/thoron progeny in the air is known or assumed. The disadvantage of this procedure is the requirement for two pumps, which leads to uncertainties due to fluctuations of air flow rate. This procedure was used by George.[3]

2. Counting the reference filter and the wire screen. This procedure allows determination of the measurement efficiency (combined collection-counting efficiency). The requirements and disadvantages of this procedure are similar to those for procedure 1.
3. Counting the reference filter, the back-up filter, and the wire screen. From these counts, the geometric counting efficiency of the screen can be determined. This procedure was used by Stranden and Berteig.[5] Additionally, the collection efficiency may be determined if the unattached fraction is known. Again, two pumps are required, with attendant disadvantages. Modifications to this procedure were suggested by James et al.[4] The relative α-counting efficiency of the wire screen and the back-up filter was determined from measurements of the 4π γ-ray emissions from ^{214}Bi (RaC) and the corresponding α– counts from ^{214}Po (RaC'). Collection efficiency was determined on the basis of a known unattached fraction. Values of the unattached fraction were provided by diffusion battery measurements.
4. Counting the wire screen and the back-up filter. This is the simplest and potentially most accurate procedure since only a single pump is required. The procedure yields the value of collection or counting efficiency provided that the value of the other efficiency may be assumed, and the unattached fraction is known.

For the present work, procedure 4 was chosen because of its simplicity and the inherent accuracy which results from use of only one pump. The value of the counting efficiency was taken to be that determined theoretically (Section 1.2.2).

The unattached fraction was determined from:[13]

$$f = \frac{A_{ws}/KE}{A_{ws}/K + A_{buf}} \tag{22}$$

where A_{ws} and A_{buf} represent the count rates (counts per minute, cpm) on the front side of the wire screen and the back-up filter respectively, K is the geometric counting efficiency, and E is the collection efficiency of the wire screen. Rearrangement of Equation 22 leads to:

$$E = \frac{1}{f}\frac{A_{ws}}{A_{ws} + KA_{buf}} \tag{23}$$

For the measurements, conditions were arranged so that the unattached fraction constituted most of the radon progeny. The attached fraction was negligible (this point is discussed later). The parameter f is then equal to unity, and the equation for collection efficiency takes the simple form:

$$E = \frac{A_{ws}}{A_{ws} + KA_{buf}} \tag{24}$$

Both the wire screen and the back-up filter were counted by the same counter (as described below) in order not to introduce additional uncertainties associated with the different efficiencies of two different counters.

Finally, the experimentally determined value of the collection efficiency may be compared with the value calculated theoretically. The agreement of these two values may serve as support for the corrections made in the theoretical models of collection efficiency and of counting efficiency. The combined efficiency may be obtained as the product KE. This product is the most important result from a practical point of view.

Experimental Apparatus

The experimental arrangement is shown schematically in Figure 4. Clean air carrying radon is introduced to a 20-l bell jar through a system of filters and silica gel column. The flow rate of the air was controlled in the range between 10 and 20 L min^{-1}. The age of the air in the jar was therefore between 1.9 and 0.85 min. The outlet of the jar was vented to a fumehood through an exhaust line. There were two other air outlets, one to the Condensation Nuclei Counter (Environment One) and one to the wire screen. The filter holder for the wire screen and the back-up filter was mounted directly in the rubber bung of the jar. The wire screen was held 3 mm away from the filter (Millipore, type AA) by means of an O-ring. The air flow through the wire screen was effected by a Dupont pump at rates in the range 1.45-2.85 L min^{-1}.

Before starting the measurements, the radon-laden air was allowed to flow through the jar at a constant rate for about half an hour. When the conditions in the jar had stabilized, sampling of the air was carried out for 5 min, and, after waiting for 45 s, counting of both wire screen (front and rear-face) and back-up filter was carried out in sequence for 2 min (with a 15 s interval between each counting) using a single Trimet alpha counter. The correction for growth of ^{214}Bi activity between the beginning and end of the measuring period can be calculated as about 1.5% according to the procedure of Wojcik and Morawska.[14] This correction of 1.5% in the raw counts is not taken into account. Neglect of this correction would lead to a final uncertainty of <0.5% in the experimental estimate of the collection-counting efficiency of the screen. Each raw count is corrected for background and for ^{218}Po decay between the beginning and the end of the counting period.

The procedure adopted for the measurements was aimed at achieving conditions under which the unattached fraction is close to unity. A very high unattached fraction is possible because the air entering the system has a very low aerosol

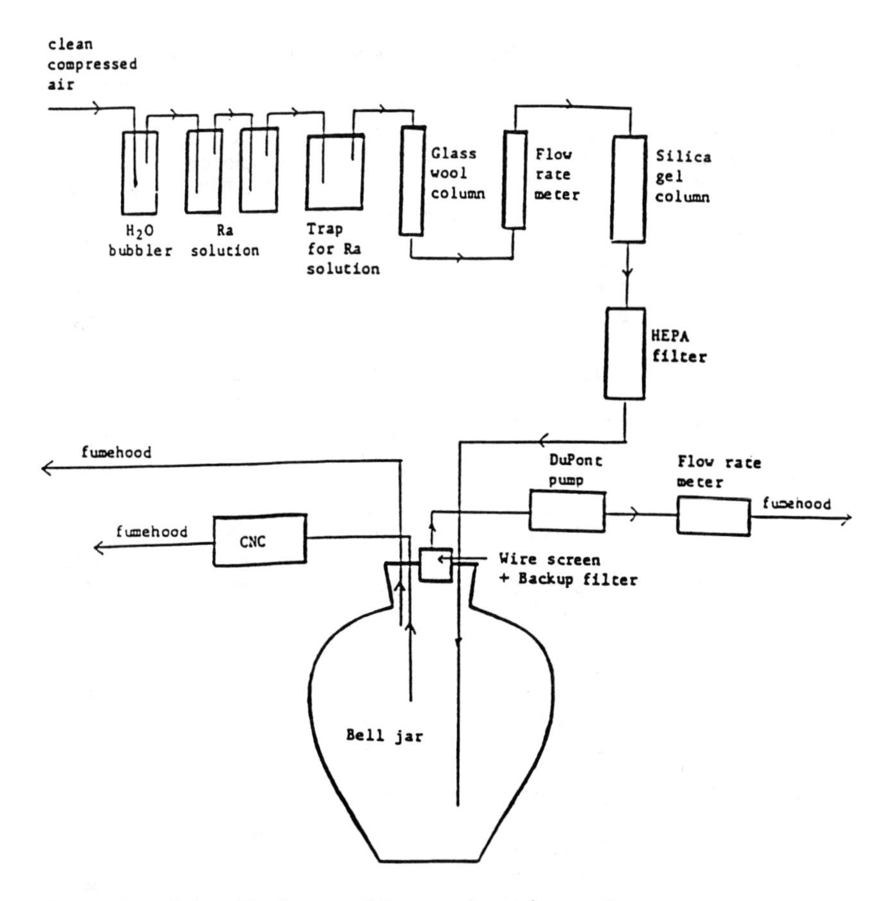

Figure 4. Schematic diagram of the experimental apparatus.

concentration (~1000 particles cm^{-3}) and because its residence time is short, so that the radon progeny in the system do not have time to attach to any particles present. The fact that the air in the jar is relatively young guarantees that almost all of the activity (>99%) on the wire screen and the back-up filter at the moment of measurement is due to ^{218}Po.

Results and Discussion

The results of the measurements are presented in Table 5. For all the measurements, the concentration of condensation nuclei in the air was about 10^3 cm^{-3}. This quality of air was achieved by use of a HEPA filter (Gelman Sciences) at the air inlet. Millipore filters (AA 0.45 μm) were tried in an attempt to reduce the condensation nuclei count even lower, but the resistance to air flow was unacceptable.

In order to check whether, at the condensation nuclei concentration of 10^3 cm^{-3}, it was reasonable to assume that the unattached fraction f was unity, measurements were carried out at increasing air ages.

Table 5. Results of Laboratory Measurements for 60 Mesh Tyler Wire Screen

Age of the Air (min)	Face Velocity (cm s^{-1})	Ratio[a] $N_1:N_2:N_3$	Ratio N_1/N_2	Collection[b] Efficiency $K_1 = 0.81$	$K_2 = 0.88$
3.00	10.41	4.26:1.17:1.0	3.64	0.84	0.83
2.30	10.41	4.97:1.49:1.0	3.33	0.86	0.85
1.90	7.70	1.30:2.19:1.0	2.54	0.93	0.93
		4.84:1.56:1.0	3.11	0.86	0.85
		6.39:1.64:1.0	3.89	0.89	0.88
		9.76:2.22:1.0	4.40	0.92	0.92
				<0.90>[c]	<0.89>
	10.41	4.21:1.28:1.0	3.29	0.84	0.83
		4.10:3.48:1.0	3.70	0.84	0.82
		4.72:1.38:1.0	3.42	0.85	0.84
		4.28:1.34:1.0	3.19	0.84	0.83
		3.33:1.03:1.0	3.24	0.80	0.79
				<0.83>	<0.82>
	13.18	3.28:0.95:1.0	3.45	0.80	0.79
		4.14:1.34:1.0	3.09	0.84	0.82
		3.74:0.94:1.0	3.98	0.82	0.81
		5.02:1.34:1.0	3.75	0.86	0.85
				<0.83>	<0.82>
1.58	7.70	8.83:1.95:1.0	4.52	0.92	0.91
		7.45:1.80:1.0	4.13	0.90	0.89
		8.14:1.79:1.0	4.55	0.91	0.90
		8.52:2.45:1.0	3.47	0.91	0.91
				<0.91>	<0.90>
	10.41	7.72:2.53:1.0	3.05	0.91	0.90
		5.35:1.19:1.0	4.50	0.87	0.86
		4.75:1.47:1.0	3.24	0.85	0.84
		5.95:1.29:1.0	4.61	0.88	0.87
		6.15:1.57:1.0	3.92	0.88	0.87
				<0.88>	<0.87>
	13.18	4.06:0.98:1.0	4.15	0.83	0.82
		4.10:1.06:1.0	3.86	0.84	0.82
		4.28:1.35:1.0	3.17	0.84	0.83
		3.58:0.87:1.0	4.12	0.82	0.80
1.27	7.70	10.11:2.74:1.0	4.02	0.93	0.92
		9.82:3.09:1.0	3.17	0.92	0.92
		9.55:2.49:1.0	3.83	0.92	0.92
		7.93:1.72:1.0	4.61	0.91	0.90
				<0.92>	<0.91>
	10.41	5.40:1.86:1.0	2.91	0.87	0.86
		5.53:1.97:1.0	2.81	0.87	0.86
		4.76:1.71:1.0	2.79	0.85	0.84
		6.31:1.86:1.0	3.39	0.89	0.89
				<0.87>	<0.86>
	13.18	4.80:1.73:1.0	2.78	0.86	0.85
		3.46:1.09:1.0	3.17	0.81	0.80

Table 5. Results of Laboratory Measurements for 60 Mesh Tyler Wire Screen (continued)

Age of the Air (min)	Face Velocity (cm s^{-1})	Ratio[a] $N_1:N_2:N_3$	Ratio $\frac{N_1}{N_2}$	Collection[b] Efficiency $K_1 = 0.81$	$K_2 = 0.88$
		4.40:1.59:1.0	2.78	0.84	0.83
		5.16:1.48:1.0	3.48	0.86	0.85
				<0.84>	<0.83>
0.95	7.70	4.47:1.18:1.0	3.79	0.85	0.84
		6.31:2.36:1.0	2.67	0.89	0.88
		7.33:1.15:1.0	6.37	0.90	0.90
		9.88:3.05:1.0	3.24	0.92	0.92
				<0.89>	<0.89>
	10.41	4.40:1.33:1.0	3.31	0.84	0.83
		13.50:3.07:1.0	4.40	0.94	0.94
		4.38:0.93:1.0	4.72	0.84	0.83
		6.93:1.53:1.0	4.52	0.90	0.89
				<0.88>	<0.87>
	13.18	6.91:2.39:1.0	2.89	0.90	0.89
		4.20:1.53:1.0	2.75	0.84	0.83
		3.17:0.98:1.0	3.24	0.80	0.78
		4.62:1.23:1.0	3.76	0.85	0.84
				<0.85>	<0.83>
	14.84	2.60:0.62:1.0	4.20	0.76	0.75

[a] This ratio is the ratio of counts after correction for background and for the decay of ^{218}Po.
[b] $E = N_1/(N_1 + KN_3)$ where K = geometric counting efficiency.
[c] < > = Average of columns above.

In columns five and six of Table 5 are presented the computed values for collection efficiency of the wire screen for values of the counting efficiency K_1 = 0.81 and K_2 = 0.88, respectively. K_1 is the theoretically calculated value (Section 1.2.2). The value of K_2 is discussed later.

Average collection efficiencies are shown in Table 6 as a function of the age of the air and the face velocity.

Calculations were performed on the assumption that there is no attached fraction present in the air, so that the parameter f in Equation 2 is equal to unity. However, if this assumption is not true, and there is an attached fraction, parameter f would be <1.0 and the determined value of the collection efficiency would be an underestimate. In such a case, with decrease of the age of the air, the attached fraction would be smaller, the value of f closer to unity, and the calculated collection efficiency higher. In summary, any observed increase in efficiency with decrease in the age of the air would suggest that the assumption of zero attached fraction in the air is not true. Inspection of Table 6 shows no such increase. In

Table 6. Dependence of the Collection Efficiency on the Age of the Air and the Face Velocity (K_1 = 0.81)[a] for 60 Mesh Wire Screen

Face Velocity (cm s^{-1})	Age of the Air (min)					
	3.00	2.30	1.90	1.58	1.27	0.95
7.70			0.90	0.91	0.92	0.89 (0.91)[b]
10.41	0.84	0.86	0.83	0.88	0.87	0.88 (0.86)
13.18			0.83	0.84	0.85	0.85 (0.84)
14.84						0.76 (0.76)

[a] K_1 = geometric efficiency.
[b] ( ) = average for each face velocity.

support of this conclusion, calculations based on a theoretical attachment model (Phillips et al. 1988) give the following results:

1. For an aerosol particle radius of 0.035 μ, (R_1), the difference in collection efficiency, E, between ages of 1 and 2 min is 1%.
2. For an aerosol particle radius of 0.05 μ (R_2), the difference in collection efficiency, E, between ages of 1 and 2 min is 3%. In no measurement was a 3% difference between 1 and 2 min observed, suggesting that the AMAD is closer to 0.035 μ (R_1) (or even smaller) than 0.05 μm (R_2) and that the collection efficiency is underestimated by less than 1%. Any error arising from the presence of attached fraction is of a systematic type (one direction, not ±). Accordingly, the attached fraction may be deemed to be negligible.

The decrease in collection efficiency with increase in face velocity is predicted by theoretical calculations and observed in earlier research.[3,4] The average collection efficiencies for face velocities 7.70, 10.41, and 13.18 cm/s are 0.91, 0.86, and 0.84; the theoretically calculated values are 0.88, 0.81, and 0.75, respectively. Thus, the experimental values are slightly higher than the theoretical values. Likely reasons for the small differences lie in the assumptions needed for the theoretical calculation. First, the geometry of the screen was simplified. Second, uniform deposition was assumed around each wire. In calculating the average value of the view factor for the upper and lower half of the screen, the sum was divided by two, which means that the contribution of each half, because of the uniform distribution, is 0.5.

From the results presented in column four of Table 5, the average front-to-back ratio for α counts on the wire screen is 3.64. Holub and Knutson obtained a value of 2.5 for a face velocity of 18 to 21 cm/s (rather than the face velocity used here of 11.5 cm/s).[12] The higher value of 3.64 found here is consistent with the lower face velocity. If it is assumed that the ratio determined experimentally (3.64) reflects the difference in deposition between the upper and lower halves of the screen, it is possible to calculate, instead of the average view factor, the weighted view factor, as (0.467)(0.784) + (0.344)(0.216) = 0.440, where 0.467 and 0.344

Table 7. Dependence of the Collection Efficiency and the Total Measurement Efficiency (Alternatively Termed the Combined Collection-Counting Efficiency), KE, on the face velocity (K_2 = 0.88)[a] for 60 Mesh Wire Screen

Face Velocity (cm s^{-1})	Collection Efficiency (%)	Measurement Efficiency (%)
7.70	0.90 + 0.01	0.79
10.41	0.85 + 0.02	0.75
13.18	0.83 + 0.01	0.73
14.48	0.75 + 0.02	0.66

[a] K_2 = geometric efficiency.

are the previously calculated geometrical view factors for the upper and the lower half of the screen and 0.784 and 0.216 (calculated from the ratio 3.64) represent the contributions of the upper and the lower half, respectively, based on the experimentally observed nonuniform deposition. For comparison purposes, it should be noted that the view factor of the filter is 0.50.

The weighted view factor of 0.44 (K_2 = 0.88) is valid for the front face wire screen measurements. In determination of efficiency, as well as in use of the wire screen to measure unattached fraction, the front face is always counted.

As expected, the counting efficiency increases if the weighted view factor is increased. Results based on the weighted view factor are presented in column 6 in Table 5 and in Table 7. Also shown in Table 7 are values of the total measurement efficiency, KE, where K is the geometric efficiency and E is the experimental collection efficiency for various face velocities. Comparison of the collection efficiency values 0.85 ± 0.02 (relative standard deviation) and 0.83 ± 0.01 (relative standard deviation) for face velocities 10.41 and 13.18 cm/s with the theoretically calculated values 0.81 and 0.75 shows that the theoretical values are still slightly lower than the experimental values. The theoretical values are of course dependent on the simplifying assumptions used in the calculations, as noted earlier.

Summary of 60 Mesh Tyler Wire Screen Results

The experiments to determine the combined collection-counting efficiency of wire screens were carried out under the conditions used for the theoretical calculation. The measurements were performed using air containing a negligible attached fraction. The average experimental collection efficiencies for face velocities 7.70, 10.41, and 13.18 cm/s are 0.90, 0.85, and 0.83, respectively. The experimental values are slightly higher than the theoretical values (0.88, 0.81, and 0.75, respectively), the most likely reason being the assumptions made for the theoretical calculations.

The combined collection-counting efficiencies based on experimental collection efficiencies and theoretical geometric efficiencies (measurement efficiency) are 0.79, 0.75, and 0.73 for face velocities 7.70, 10.47, and 13.18 cm/s, respectively. For practical application, these values of the measurement efficiency for a 60 mesh per inch Tyler wire screen are recommended.

Table 8. Wire Screen Dimensions

Tyler Mesh Size	Wire Diameter (mm)	Size of Opening (mm)
35	0.343	0.383
40	0.254	0.381
50	0.228	0.280
60	0.178	0.250
80	0.140	0.177
100	0.114	0.140
120	0.094	0.118

Table 9. Geometric Counting Efficiency for Wire Screens

Tyler Mesh Size	View Factor of the Upper Part of the Screen	View Factor of the Lower Part of the Screen	Geometric Counting Efficiency	Corrected Geometric Counting Efficiency
35	0.4754	0.3270	0.80	0.89
40	0.4854	0.3634	0.85	0.92
50	0.4660	0.3305	0.80	0.87
60	0.4670	0.3437	0.81	0.88
80	0.4520	0.3234	0.78	0.85
100	0.4454	0.3158	0.76	0.83
120	0.4435	0.3166	0.76	0.83

DETERMINATION OF THE COMBINED COLLECTION-COUNTING EFFICIENCY OF A WIRE SCREEN: EXTENSION TO SEVEN DIFFERENT WIRE SCREEN SIZES

For seven different wire screen sizes, the collection and counting data computed on the basis of the theoretical model are next compared with experimental results obtained using two different methods. The theoretical calculations and the experimental measurements were performed for the 35, 40, 50, 60, 80, and 120 Tyler mesh screens, and three different face velocities, namely, 7.70, 10.41, and 13.18 cm s^{-1}. The outcome of the work is the determination of recommended values of the combined collection-counting efficiencies for these screens.

Theoretical Determination of the Collection and Counting Efficiency

Determination of Counting Efficiency

Counting efficiency for all seven wire screens were calculated according to the theoretical model discussed in Section 1.2.1. The parameters of the screens, wire diameters and size of openings are given in Table 8.

The geometric counting efficiencies for all wire screens are presented in Table 9.

Table 10. Interstitial Velocity (cm/s) as a Function of Face Velocity for Various Tyler Mesh Screens

	Face Velocity [cm/s]		
Tyler Mesh Size	7.70	10.41	13.18
35	17.68	23.90	30.26
40	14.55	19.66	24.90
50	16.52	22.34	28.28
60	15.13	20.46	25.90
80	16.18	21.87	27.69
100	16.52	22.34	28.28
120	16.28	22.01	27.86

Table 11. Collection Efficiency and Combined Collection-Counting Efficiency Calculated on the Basis of the Theoretical Model

	Face Velocity [cm/s]					
	7.70		10.41		13.18	
Tyler Mesh Size	Collection Efficiency	Combined Collection-Counting Efficiency	Collection Efficiency	Combined Collection-Counting Efficiency	Collection Efficiency	Combined Collection-Counting Efficiency
35	0.79	0.63	0.71	0.57	0.65	0.52
40	0.79	0.67	0.71	0.60	0.65	0.55
50	0.86	0.69	0.78	0.62	0.72	0.58
60	0.88	0.71	0.81	0.66	0.75	0.61
80	0.93	0.73	0.87	0.68	0.82	0.64
100	0.96	0.73	0.92	0.70	0.87	0.66
120	0.98	0.74	0.94	0.71	0.90	0.68

Theoretical Calculations of Collection Efficiency

Theoretical calculations of collection efficiencies for the wire screens, were performed according to the method described in Section 1.2.3. The results of collection efficiency calculations for all wire screens for three different values of face velocities are listed in Table 11. Values shown are those of the experimental face velocity. The interstitial velocity for each wire screen and each face velocity was calculated according to the example given for the 60 mesh wire screen. Interstitial velocities are listed in Table 10. For these calculations, the diffusion coefficient was taken as 0.054 cm^2/s.

The overall efficiencies (combined collection-counting efficiencies) based on the theoretical models are also shown in Table 11.

Experimental Determination of Combined Collection-Counting Efficiency of Wire Screens

For the present work, two experimental methods were chosen. The method in which the wire screen and the back-up-filter are counted is considered the best method. This method was discussed in detail in Section 1.3.1. In order to provide a cross-check, a method based on counting and reference filter and the wire screen was also used. In this method, the combined collection-counting efficiency is determined on a purely experimental basis. According to the latter method, the unattached fraction is determined from:

$$f = \frac{A_{ws}}{KE \times A_{ref}} \quad (25)$$

where A_{ws} represents, as previously, the count rate on the front side of the wire screen, and A_{ref} is the count rate of the reference filter. For f = 1.0, Equation 25 takes the form:

$$KE = \frac{A_{ws}}{A_{ref}} \quad (26)$$

As mentioned previously, the main disadvantage of this last method is the need to use two different pumps or to perform two separate measurements. As a consequence, the accuracy of the resultant value of KE is poorer than the accuracy of the value of E determined according to the method based on counting the wire screen and the back-up filter. However, since the resultant value of KE is purely experimental, it may usefully be compared with the value of KE derived by the partly theoretical and partly experimental method.

The measurement conditions for all the experiments were the same as in Section 1. The age of the air for the measurements was 1.12 min. The experiments to determine the combined collection-counting efficiency of wire screens were carried out under conditions as assumed for the theoretical calculation. The measurements were performed with air containing a negligible attached fraction (<3%).

The results for all wire screens are presented in Table 12.

The average front-to-back ratio for alpha counts for all the wire screens is 3.59, which is in good agreement with the value 3.64, determined earlier (Section 1). The value of 3.59 was used as the basis for correcting the geometric efficiency. The correction was performed in the same way as for the 60 mesh screen. The corrected values of geometric counting efficiency are given in the last column of Table 9. Results for efficiency based on corrected values of the geometric counting efficiency are given in column four of Table 12. Results for the efficiency of the seven wire screens investigated are shown in Tables 13 to 17. Results for the collection efficiency for both uncorrected and corrected values of K are shown in Tables 13 and 14. In Table 16, the combined efficiency is calculated with the product of KE for both uncorrected and corrected values of K.

From Tables 13 and 14, it is apparent that for each wire screen the collection efficiency decreases with increasing values of face velocity. Also, for each face

Table 12. Results of Laboratory Measurements

Tyler Mesh Size	Face Velocity ($cm\ s^{-1}$)	Ratio[a] $N_1:N_2:N_3$	Ratio N_1/N_2	Collection[b] Efficiency		Combined Efficiency
				K_1=0.81	K_2=0.88	
35 Mesh	7.70					
		7.05:2.15:1.0	3.27	0.90	0.89	1.09
		5.83:1.96:1.0	2.98	0.88	0.87	0.88
		4.66:0.68:1.0	6.84	0.85	0.84	0.81
		4.38:1.57:1.0	2.79	0.85	0.83	0.72
				<0.87>[c]	<0.86>	
	10.41					
		3.07:0.99:1.0	3.10	0.79	0.78	0.49
		3.88:2.12:1.0	1.83	0.83	0.81	0.57
		7.27:2.98:1.0	2.44	0.90	0.89	0.96
		2.30:0.72:1.0	3.17	0.74	0.72	0.78
				<0.82>	<0.80>	
	13.18					
		2.48:1.13:1.0	2.20	0.76	0.74	0.55
		1.91:1.13:1.0	1.69	0.70	0.68	0.33
		1.88:1.19:1.0	1.58	0.70	0.68	0.38
				<0.72>	<0.70>	
40 Mesh						
	7.70					
		4.21:1.17:1.0	3.61	0.83	0.82	0.69
		4.73:1.09:1.0	4.34	0.85	0.84	0.71
		5.00:1.76:1.0	2.84	0.85	0.85	0.47
		3.26:1.19:1.0	2.73	0.79	0.78	0.61
				<0.83>	<0.82>	
	10.41					
		2.46:1.30:1.0	1.90	0.74	0.73	0.49
		3.26:1.20:1.0	2.72	0.79	0.78	0.68
		2.81:0.96:1.0	2.94	0.77	0.75	0.78
		2.48:1.09:1.0	2.28	0.74	0.73	0.43
				<0.76>	<0.75>	
	13.18					
		2.48:0.95:1.0	2.61	0.75	0.73	0.45
		2.20:0.62:1.0	3.56	0.72	0.71	0.62
		2.42:0.68:1.0	3.58	0.74	0.73	0.68
				<0.74>	<0.72>	
50 Mesh						
	7.70					
		6.90:0.99:1.0	6.97	0.90	0.89	0.47
		16.78:6.35:1.0	2.64	0.95	0.95	0.76
		7.43:1.87:1.0	3.97	0.90	0.89	0.80
		4.99:0.93:1.0	5.37	0.86	0.85	0.72
				<0.90>	<0.90>	
	10.41					
		5.38:2.96:1.0	1.82	0.87	0.86	0.67
		5.47:1.32:1.0	4.14	0.87	0.86	0.92
		6.81:1.19:1.0	5.71	0.89	0.89	0.53
		5.91:2.04:1.0	2.90	0.88	0.87	0.75
				<0.88>	<0.87>	

Table 12. Results of Laboratory Measurements (continued)

Tyler Mesh Size	Face Velocity (cm s^{-1})	Ratio[a] $N_1:N_2:N_3$	Ratio N^1/N^2	Collection[b] Efficiency		Combined Efficiency
				$K_1=0.81$	$K_2=0.88$	
	13.18					
		3.49:1.20:1.0	2.90	0.81	0.80	0.80
		3.97:1.30:1.0	3.06	0.83	0.82	0.69
		4.28:1.36:1.0	3.14	0.84	0.83	0.52
		3.32:1.32:1.0	2.50	0.81	0.79	0.57
				<0.82>	<0.81>	
60 Mesh						
	7.70					
		10.32:2.31:1.0	4.46	0.93	0.92	0.73
		7.68:2.46:1.0	3.12	0.90	0.90	0.67
		10.30:3.31:1.0	3.11	0.93	0.92	0.85
				<0.92>	<0.91>	
	10.41					
		6.26:1.59:1.0	3.94	0.89	0.88	0.87
		4.29:2.45:1.0	1.75	0.84	0.83	0.43
		6.90:2.24:1.0	3.08	0.89	0.89	1.79
		5.42:2.53:1.0	2.14	0.87	0.86	0.76
		6.90:1.66:1.0	4.15	0.89	0.89	0.64
		6.63:2.35:1.0	2.82	0.89	0.88	0.68
				<0.88>	<0.87>	
	13.18					
		4.40:1.82:1.0	2.42	0.84	0.83	0.61
		5.13:1.65:1.0	3.11	0.86	0.85	0.82
		5.05:1.55:1.0	3.26	0.86	0.85	0.70
		5.79:1.50:1.0	3.87	0.88	0.87	0.88
		4.22:1.63:1.0	2.58	0.84	0.83	0.83
				<0.86>	<0.85>	
80 Mesh						
	7.70					
		8.99:1.63:1.0	5.50	0.92	0.91	0.85
		12.38:3.31:1.0	3.74	0.94	0.94	1.01
		17.25:5.20:1.0	3.32	0.96	0.95	0.72
		15.37:2.72:1.0	5.64	0.95	0.95	1.18
				<0.94>	<0.94>	
	10.41					
		7.84:2.66:1.0	2.95	0.91	0.90	1.54
		7.97:3.01:1.0	2.65	0.91	0.90	1.13
		12.62:3.52:1.0	3.58	0.94	0.94	0.69
		7.23:1.30:1.0	5.58	0.90	0.90	0.58
		13.88:3.68:1.0	3.77	0.95	0.94	1.82
				<0.92>	<0.92>	
	13.18					
		5.26:2.10:1.0	2.51	0.87	0.86	0.68
		6.66:2.36:1.0	2.82	0.90	0.89	0.50
		5.22:1.22:1.0	4.27	0.87	0.86	0.71
				<0.88>	<0.87>	

Table 12. Results of Laboratory Measurements (continued)

Tyler Mesh Size	Face Velocity (cm s^{-1})	Ratio[a] $N_1:N_2:N_3$	Ratio N_1/N_2	Collection[b] Efficiency		Combined Efficiency
				K_1=0.81	K_2=0.88	
100 Mesh						
	7.70					
		16.39:4.46:1.0	3.67	0.96	0.95	1.01
		20.05:4.81:1.0	4.17	0.96	0.96	0.94
		11.35:2.84:1.0	4.00	0.94	0.93	0.86
		24.01:5.94:1.0	4.04	0.97	0.97	0.84
				<0.96>	<0.95>	
	10.41					
		9.04:1.96:1.0	4.60	0.92	0.92	1.16
		8.41:1.84:1.0	4.56	0.92	0.91	1.05
		15.27:3.02:1.0	5.06	0.95	0.95	0.94
		11.55:2.72:1.09	4.24	0.94	0.93	1.32
		6.98:2.27:1.0	3.08	0.90	0.89	0.60
		10.68:2.49:1.0	4.28	0.93	0.93	0.95
				<0.93>	<0.92>	
	13.18					
		9.90:2.90:1.0	3.42	0.93	0.92	0.94
		9.04:2.27:1.0	3.98	0.92	0.92	0.63
		10.32:2.64:1.0	3.91	0.93	0.93	0.58
		12.45:3.58:1.0	3.48	0.94	0.94	0.85
				<0.93>	<0.93>	
120 Mesh						
	7.70					
		24.31:4.82:1.0	5.05	0.97	0.97	0.86
		15.54:2.66:1.0	5.83	0.95	0.95	1.00
		14.85:2.52:1.0	5.89	0.95	0.95	0.64
		15.46:5.16:1.0	3.00	0.95	0.95	0.61
				<0.96>	<0.95>	
	10.41					
		15.68:5.67:1.0	2.76	0.95	0.95	0.63
		13.03:3.76:1.0	3.48	0.94	0.94	0.86
		9.28:2.38:1.0	3.90	0.92	0.92	1.38
		13.20:3.03:1.0	4.36	0.95	0.94	0.66
				<0.94>	<0.94>	
	13.18					
		13.75:3.99:1.0	3.44	0.95	0.94	0.78
		12.17:3.12:1.0	3.90	0.94	0.94	0.84
		12.14:2.40:1.0	5.06	0.94	0.94	0.94
		8.62:1.77:1.0	4.88	0.92	0.91	0.69
				<0.94>	<0.93>	

[a] This ratio is the ratio counts after correction for background and for the decay of ^{218}Pb.
[b] $E = N_1/(N_1 + KN_3)$ where K = geometric counting efficiency.
[c] < > Average of columns above.

Table 13. Collection Efficiency, $E = A_{ws}/(A_{ws} + KA_{buf})$, Based on Uncorrected Values of K (Basis: Theoretical Calculation)

Tyler Mesh Size	Face Velocity [cm s^{-1}] 7.70	10.41	13.18
35	0.87 ± 0.02	0.82 ± 0.07	0.72 ± 0.03
40	0.83 ± 0.03	0.76 ± 0.02	0.74 ± 0.01
50	0.90 ± 0.04	0.88 ± 0.01	0.82 ± 0.02
60	0.92 ± 0.01	0.88 ± 0.02	0.86 ± 0.02
80	0.94 ± 0.02	0.92 ± 0.02	0.88 ± 0.01
100	0.96 ± 0.01	0.93 ± 0.02	0.93 ± 0.01
120	0.96 ± 0.01	0.94 ± 0.01	0.94 ± 0.01

Table 14. Collection Efficiency, $E = A_{ws}/(A_{ws} + KA_{buf})$ based on corrected values of K (Basis: Theoretical Calculation Plus Experimental Ratio of the Front-to-Back Alpha Counts of the Wire Screen)

Tyler Mesh Size	Face Velocity [cm s^{-1}] 7.70	10.41	13.18
35	0.86 ± 0.03	0.80 ± 0.07	0.72 ± 0.03
40	0.82 ± 0.03	0.75 ± 0.02	0.72 ± 0.01
50	0.90 ± 0.04	0.87 ± 0.01	0.81 ± 0.02
60	0.91 ± 0.01	0.87 ± 0.01	0.85 ± 0.02
80	0.94 ± 0.02	0.92 ± 0.02	0.87 ± 0.02
100	0.95 ± 0.02	0.92 ± 0.02	0.93 ± 0.01
120	0.95 ± 0.01	0.94 ± 0.01	0.93 ± 0.01

velocity, the collection efficiency increases with increasing mesh number of the wire screen. This pattern is not maintained for the first two wire screens, 35 and 40 mesh; for example, for face velocity 7.7 cm/s, the collection efficiencies for these screens are 0.87 and 0.83, respectively. Such results are, however, expected, based on the relationship between wire diameter and opening size for these two screens (see Table 8). The opening size is almost the same for these screens, whereas the wire diameter of the 40 mesh screen is about one third smaller than for the 35 mesh screen.

The collection efficiency data derived on the basis of theoretical, corrected values of view factor and measurements (from Table 14) are compared with the collection efficiency data derived on the basis of theoretical deposition results (see Table 15). The data in Table 15 are calculated for a face velocity of 10.41 cm/s and for a diffusivity of 0.054 cm^2/s. The experimental collection efficiencies agree closely with the theoretical efficiencies and also — for finer mesh sizes — with the Thomas and Hinchliffe efficiencies. Despite the fact that the theoretical calculations are not fully realistic, as they do not take into account nonuniform deposition, there is still good agreement between the experimental and theoretical

Table 15. Comparison of Collection Efficiencies for Face Velocity V = 10.41 cm s^{-1} (D = 0.054 cm^2 s^{-1})

Tyler Mesh Size	Efficiency by Thomas and Hinchliffe	Efficiency by Cheng and Yeh	Efficiency by Finite-Difference	Experimental Efficiency (See Table 14)
35	0.63	0.64	0.71	0.80
40	0.62	0.64	0.71	0.75
50	0.72	0.72	0.78	0.87
60	0.75	0.81	0.81	0.87
80	0.85	0.81	0.87	0.92
100	0.90	0.83	0.92	0.92
120	0.94	0.84	0.94	0.94

Table 16. Combined Efficiency, KE, for Uncorrected and Corrected Values of K

Tyler Mesh Size	Face Velocity (cm s^{-1}) 7.70		10.41		13.18	
	Uncorrected	Corrected	Uncorrected	Corrected	Uncorrected	Corrected
35	0.70	0.76	0.65	0.71	0.58	0.62
40	0.71	0.75	0.65	0.69	0.63	0.66
50	0.72	0.78	0.70	0.76	0.66	0.71
60	0.74	0.80	0.71	0.77	0.69	0.75
80	0.74	0.79	0.72	0.78	0.69	0.74
100	0.73	0.79	0.70	0.77	0.71	0.77
120	0.73	0.79	0.72	0.78	0.71	0.78

data. For practical application, the values derived on the experimental basis (Table 14) are recommended.

Table 17 gives average values of the combined experimental efficiency. These values have been calculated from data in the last column of Table 12. In contrast to the values of collection efficiency in columns 3 and 4 of this table, the combined experimental efficiency values are much more scattered. As mentioned previously, the scatter is due to the fact that it was necessary to perform two measurements in order to obtain one value of combined efficiency: in the first measurement, air was drawn through the wire screen, and, in the second, through the reference filter. Face velocity fluctuations were caused by fluctuations in the pump flow rate between the first and the second measurement. In addition, even small fluctuations in the flow rate of air through the radon source caused noticeable changes in radon activity in the holding vessel, and, as a consequence, changes in the activity collected on the wire screen or filter. Because of these problems, some of the results obtained in this way are meaningless and were rejected from further analysis. It was decided to reject the values of combined efficiency that were greater than 1.0 or smaller than 0.5. These limits are arbitrary, but are reasonable since values outside the range 0.5 to 1.0 are not justified by the physics of the problem.

Table 17. Combined Experimental Efficiency, KE = A_{ws}/A_{ref}

Tyler Mesh Size	Face Velocity (cm s^{-1})		
	7.70	10.41	13.18
35	0.80 (3)[a]	0.70 (4)	0.55 (1)
40	0.67 (3)	0.65 (3)	0.65 (2)
50	0.76 (3)	0.72 (4)	0.65 (4)
60	0.75 (3)	0.74 (4)	0.77 (4)
80	0.79 (2)	0.64 (2)	0.63 (3)
100	0.88 (3)	0.83 (3)	0.75 (4)
120	0.78 (4)	0.72 (3)	0.81 (4)

[a] In parentheses, number of data points averaged, after rejection of physically impossible values >1.00 and <0.50.

By comparing the combined experimental efficiencies in Table 17 with those in Table 16, it can be seen that there is good agreement between the two sets of data. However, it is important to keep in mind all the problems that influence the experimental determination of combined efficiency on a purely experimental basis (procedure 2). On account of these problems, for practical applications, values of the combined efficiency derived using experimental procedure 4 and theoretical calculations of view factor are recommended for use.

Summary and Conclusions

Seven wire screens of different mesh sizes were investigated. The mesh sizes were 35, 40, 50, 60, 80, 100, and 120. As a sample of the results, the average experimental collection efficiencies determined for the 35 mesh screen at face velocities 7.7, 10.41, and 13.18 cm s^{-1} were 0.80, 0.86, and 0.70, respectively, and for the 120 mesh screen, 0.95, 0.94, and 0.93. These values may be compared with the theoretically calculated values, which were 0.79, 0.71, and 0.65, respectively for the 35 mesh screen, and 0.98, 0.94, and 0.90, respectively, for the 120 mesh screen. The experimental values differ slightly from the theoretical values, the most likely reason being the assumptions made for the theoretical calculations.

The combined collection-counting efficiency (measurement efficiency) determined via the most reliable means (counting the wire screen and the back-up filter, together with use of theoretical geometrical efficiencies and experimental front-back deposition ratios) are, for the previously mentioned face velocities, 0.76, 0.71, and 0.62 for the 35 mesh screen, and 0.79, 0.78, and 0.78 for the 120 mesh screen. For practical purposes, these values are recommended. (Values recommended for the 60 mesh screen are given on page 231.

ACKNOWLEDGMENTS

This work was supported by the Atomic Energy Control Board of Canada and by the Natural Sciences and Engineering Research Council of Canada.

REFERENCES

1. Thomas, J.W. and Hinchliffe, L.E., "Filtration of 0.001 μm Particles by Wire Screens," *J. Aerosol Sci.*, 3:387–393 (1972).
2. Cheng, Y.S. and Yeh, H.C., "Theory of Screen Type Diffusion Battery," *J. Aerosol Sci.*, 11:333–319 (1980).
3. George, A.C., "Measurements of Uncombined Fraction of Radon Daughters with Wire Screens," *Health Phys.*, 23:390–392 (1972).
4. James, A.C., Bradford, G.F., and Howell, D.M., "Collection of Unattached RaA Atoms Using a Wire Gauze," *Aerosol Sci.* 3:243–254 (1972).
5. Stranden, E. and Berteig, L., "Radon Daughter Equilibrium and Unattached Fraction in Mine Atmospheres," *Health Phys.*, 42:479 (1982).
6. Raghavayya, M. and Jones, J.H., "A Wire Screen-Filter Paper Combination for the Measurement of Fractions of Unattached Radon Daughters in Uranium Mines," *Health Phys.*, 26:417–429 (1974).
7. Ramamurthi, M. and Hopke, P.K., "On Improving the Validity of Wire Screen 'Unattached' Fraction Rn Daughter Measurements," *Health Phys.*, 56:189–194 (1989).
8. Cheney, W. and Kincaid, D., *Numerical Mathematics and Computing,* 2nd ed., (Monterey, CA: Brooks/Cole Publishing Company, 1985).
9. Press, W.H., Flannery, B.P., Teukolsky, S.A., and Vetterling, W.T., *Numerical Recipes. The Art of Scientific Computing,* (New York: Cambridge University Press, 1986).
10. Roache, P.J., *Computational Fluid Dynamics,* (Albuquerque: Hermosa, 1976).
11. van der Vooren, A.W. and Phillips, C.R., Report to Atomic Energy of Canada Limited, (Pinawa, 1985).
12. Holub, R.F. and Knutson, E.O., "Measuring Po-218 Diffusion Coefficient Spectra using Multiple Wire Screens," *Radon and Its Decay Products,* Hopke, P.K., (Ed.), (Washington, DC: Proc. ACS Symp. Series, 1987), pp. 340–356.
13. Khan, A., Phillips, C.R., and Dupont, P., "Analysis of Errors in the Measurement of Unattached Fractions of Radon and Thoron Progeny in a Canadian Uranium Mine Using Wire Screen Methods," *Radiat. Prot. Dosimetry,* 18(4):197–207 (1987).
14. Wojcik, M. and Morawska, L., "Radon Concentration and Exhalation Measurement with a Semiconductor Detector and an Electrostatic Precipitator Working in a Closed Circulation System," *Nucl. Instrum. Methods,* 212:393–402 (1983).
    Mercer, T.T., "Unattached Radon Decay Products in Uranium Mine Air," *Health Phys.*, 28:158 (1975).
15. Phillips, C.R., Khan, A., and Leung, H., "The Nature and Measurement of the Unattached Fraction of Radon and Thoron Progeny," in *Radon and Its Decay Products in Indoor Air,* Nazaroff, W.W. and Nero, A.V., (Eds.), (New York: John Wiley & Sons, 1988), pp. 203–256.

APPENDIX

Charlene W. Bayer
Environmental Sciences and
Technology Laboratory
Georgia Tech Research Institute
Atlanta, Georgia

A.C. Bennet
Dames & Moore
Bethesda, Maryland

Peter W.H. Binnie
Vice President
Healthy Buildings International, Inc.
Democracy Lane
Fairfax, Virginia

William W.-C. Chan
Dept. of Chemical Engineering and
Applied Chemistry
University of Toronto
Toronto, Ontario

A. Cooper
Manager
Biospherics, Inc.
Beltsville, Maryland

D.T. Dyjack
Dames & Moore
Bethesda, Maryland

D.S. Ensor
Research Triangle Institute
Research Triangle Park,
North Carolina

J.T. Hanley
Research Triangle Institute
Research Triangle Park,
North Carolina

Philip K. Hopke
Department of Chemistry
Clarkson University
Potsdam, New York

I.-C. Hsieh
Dept. of Chemical Engineering and
Applied Chemistry
University of Toronto
Toronto, Ontario

Jack G. Kay
Department of Chemistry
Drexel University
Philadelphia, Pennsylvania

George E. Keller
Union Carbide Chemicals and
Plastics Company Inc.
South Charleston, West Virginia

Richard A. Kemper
Kemper Research Foundation
Milford, Ohio

A. Khan
Dept. of Chemical Engineering and
Applied Chemistry
University of Toronto
Toronto, Ontario

Laurence S. Kirsch, Esq.
Cadawalader, Wickersham & Taft
Washington, D.C.
Indoor Pollution Law Report

S.R. Lamb
Industrial Hygienist
Biospherics, Inc.
Beltsville, Maryland

E.N. Light
Senior Scientist
Dames & Moore
Bethesda, Maryland

K.L. Long
Industrial Hygienist
Biospherics, Inc.
Beltsville, Maryland

Tracie L. Lopez
Department of Physics and
Geophysical Research Center
New Mexico Institute of
Mining and Technology
Socorro, New Mexico

Jay F. Miller
Union Carbide Chemicals and
Plastics Company Inc.
South Charleston, West Virginia

L. Morawska
Dept. of Chemical Engineering and
Applied Chemistry
University of Toronto
Toronto, Ontario
(on leave from Institue of Physics and Nuclear Techniques, The Academy of Mining and Metallurgy Kraków, Poland)

Datta V. Naik
Department of Chemistry and Physics
Monmouth College
West Long Beach, New Jersey

E. Pellizzari
Research Triangle Institute
Research Triangle Park,
North Carolina

Colin R. Phillips
Dept. of Chemical Engineering and
Applied Chemistry
University of Toronto
Toronto, Ontario

K. Ramanathan
Research Triangle Institute
Research Triangle Park,
North Carolina

Stephen D. Schery
Department of Physics and
Geophysical Research Center
New Mexico Institute of
Mining and Technology
Socorro, New Mexico

Linda S. Sheldon
Research Triangle Institute
Research Triangle Park,
North Carolina

Helen C. Shields
Bell Communications Research
Red Bank, New Jersey

J. Sickles
Research Triangle Institute
Research Triangle Park,
North Carolina

D.D. Smith
Research Triangle Institute
Research Triangle Park,
North Carolina

L.E. Sparks
U.S. Environmental Protection
Agency/
Air and Energy Engineering Research
Laboratory
Research Triangle Park,
North Carolina

A.S. Viner
Research Triangle Institute
Research Triangle Park,
North Carolina

Charles J. Weschler
Bell Communications Research
Red Bank, New Jersey

D. Westerdahl
California Air Resources Board
Sacramento, California

W. Curtis White
Dow Corning Corporation
Midland, Michigan

D. Whitaker
Research Triangle Institute
Research Triangle Park,
North Carolina

R. Wiener
U.S. Environmental Protection
Agency/
AREAL
Research Triangle Park,
North Carolina

C.S. Yang
Director
P & K Microbiology Service

Index